AF361969

Application of Climatic Data in Hydrologic Models

Application of Climatic Data in Hydrologic Models

Editors

Mohammad Valipour
Sayed M. Bateni

MDPI • Basel • Beijing • Wuhan • Barcelona • Belgrade • Manchester • Tokyo • Cluj • Tianjin

Editors
Mohammad Valipour
Metropolitan State University of Denver
USA

Sayed M. Bateni
University of Hawaii at Manoa
USA

Editorial Office
MDPI
St. Alban-Anlage 66
4052 Basel, Switzerland

This is a reprint of articles from the Special Issue published online in the open access journal *Climate* (ISSN 2225-1154) (available at: https://www.mdpi.com/journal/climate/special_issues/ climatic_data_hydrologic_models).

For citation purposes, cite each article independently as indicated on the article page online and as indicated below:

LastName, A.A.; LastName, B.B.; LastName, C.C. Article Title. *Journal Name* **Year**, *Volume Number*, Page Range.

ISBN 978-3-0365-5065-7 (Hbk)
ISBN 978-3-0365-5066-4 (PDF)

Cover image courtesy of Dr. Mohammad Valipour

Contents

Editorial

Global Surface Temperature: A New Insight

Mohammad Valipour [1], Sayed M. Bateni [1] and Changhyun Jun [2,*]

[1] Department of Civil and Environmental Engineering and Water Resources Research Center, University of Hawaii at Manoa, Honolulu, HI 96822, USA; valipour@hawaii.edu (M.V.); smbateni@hawaii.edu (S.M.B.)
[2] Department of Civil and Environmental Engineering, College of Engineering, Chung-Ang University, Seoul 06974, Korea
* Correspondence: cjun@cau.ac.kr

Citation: Valipour, M.; Bateni, S.M.; Jun, C. Global Surface Temperature: A New Insight. *Climate* **2021**, *9*, 81. https://doi.org/10.3390/cli9050081

Received: 27 April 2021
Accepted: 7 May 2021
Published: 12 May 2021

Publisher's Note: MDPI stays neutral with regard to jurisdictional claims in published maps and institutional affiliations.

This paper belongs to our Special Issue "Application of Climate Data in Hydrologic Models". Here, we represent an example of the importance of climate data in terms of global surface temperature. This paper investigates the changes of 20-year (2000–2019) mean surface temperature (ST), wind speed (WS), and albedo (AL) data from the Global Land Data Assimilation System (GLDAS) over the globe with respect to those in 1961–1990. Moreover, we assess if the alterations in ST are affected by the changes in WS and AL. We also discuss the main reasons for the variations observed in ST, WS, and AL on global and hemispheric scales.

Our planet is within ~1 °C of its highest surface temperature in the past million years [1]. The most important indicator of global warming and climate change is the global mean surface temperature (GMST) [2]. Figure 1a shows the time series of monthly GMST anomalies from the Hadly Center/Climate Research Unit (HadCRUT) [2,3]. The mean of monthly GMST anomalies in 2000–2019 is 0.54 °C higher than that in 1961–1990 (Figure 1a) (see also [1,4]). The standard deviation (SD) of the decadal variation of GMST anomalies is reduced by 0.05 °C in 2000–2019 compared to 1850–1899 (Figure 1a). This is due to the climate sensitivity to oscillation indices, particularly El Niño/Southern Oscillation (ENSO) [5]. Indeed, there were more ENSO events in 1850–1900 than in 2000–2020, which affected the GMST anomalies. For example, the high positive GMST anomaly in 1878 (shown by the black circle) is caused by the late 1870s El Niño (Figure 1a) [1].

Figure 1b shows the variation of Global Land Data Assimilation System (GLDAS) ST data (ΔST) from 1961–1990 to 2000–2019 over the globe. The results show the GMST increase of 0.66 °C from 1961–1990 to 2000–2019. Analogously, Hansen et al. [1] reported the GMST growth of ~0.2 °C per decade during the last three decades. As indicated in Figure 1b, in general, the Northern Hemisphere (NH) has a higher variation of ST (ΔST) than the Southern Hemisphere (SH) [4]. The averages of ΔST in the NH and SH are 1.27 °C and −0.18 °C, respectively. Increasing greenhouse gas (GHG) emissions and variations of the North Atlantic Oscillation (NAO) are the main causes of increasing ST across the globe and particularly in the NH [1,6]. Effective policies must focus on using clean energies to reduce the GHG emissions. Such policies can facilitate achieving the 2015 Paris COP21 Agreement to maintain the GMST below 1.5–2 °C above pre-industrial levels. The decreasing ST in the SH is associated with the variations of the SH Annular Mode (SAM) and intensification of the South Pacific Anticyclone [7,8]. Fogt and Marshall [7] showed that adiabatic cooling over the Antarctica due to the variations of SAM aids in decreasing ST in the SH.

Figure 1c indicates the boxplot of GMST anomalies in each season. The horizontal line within the box indicates the median. The upper and lower edges of the box represent the 75% and 25% percentiles, respectively. The upper and lower ends of the whiskers show the maximum and minimum values, respectively. Outliers are observations beyond the end of the whiskers. As shown, the variation of the GMST anomaly during December–February (DJF) is more than that of other seasons. Vose et al. [9] also observed the highest upward decadal trend of ST in the NH (particularly high latitudes) in DJF when the minimum ST often occurs. In addition, the global minimum ST has increased approximately twice as fast as the global maximum ST since 1950 [9]. Hence, a larger variation in the GMST anomaly is seen in DJF (Figure 1c).

Comparison of Figure 1b,d illustrates a larger variation of ST during DJF in both hemispheres. During DJF, the mean of ΔST in the NH is 1.8 °C (Figure 1d). However, in some parts of Eastern Russia, it can reach up to 20 °C (Figure 1d). Hansen et al. [1] concluded that the global warming of more than ~1 °C could lead to a dangerous climate change and impacts the sea water level.

From 1961–1990 to 2000–2019, the GLDAS ST (WS) increased (decreased) in the northwest of North America (Alaska) and northeast of Asia (Russia) (Figure 1b,e). Both ST and WS increased in Canada (except in the western part), most regions of Australia and North Africa. In most regions of Africa and South America, WS enhanced, but ST reduced. Figure 1f shows the changes in the ST-WS correlation ($\Delta\rho_{ST-WS}$) from 1961–1990 to 2000–2019. High positive or negative values of $\Delta\rho_{ST-WS}$ (>1 or <−1) in Southern Africa, North Greenland, Western America, Eastern Russia, and Central and Eastern Asia denote the changes of vegetation cover [10].

As can be seen in Figure 1e, winds slow down in most of the coastal regions as well as in Central and Eastern of Asia due to an increase in the surface roughness. These findings are consistent with those of Vautard et al. [10] who reported the highest downward trend of WS in Central and Eastern Asia from 1979 to 2008 due to the land-use change and/or biomass increase. They demonstrated that increasing the Normalized Difference Vegetation Index (NDVI) in Eurasia could explain 25–60% of the NH atmospheric stilling.

The increase of AL plays an important role in future climate change [11]. Ghimire et al. [12] reported an increase of about 1% in the global albedo due to the land cover change during 1700–2005. That caused the global atmospheric radiation to vary by almost 0.15 W/m^2. Figure 1g represents changes in the GLDAS AL from 1961–1999 to 2000–2019 over the globe. On average, AL increases by 3.19% in the NH. AL varies with changes in cloud fractional coverage, aerosol amount, and land cover type [13]. If the variation of AL is caused by changes in aerosols and land cover type, ST reduces by rising AL [13]. In the NH, AL increased from 1961–1999 to 2000–2019 (Figure 1g), while ST did not decrease (Figure 1b). Therefore, the increase of AL is indicative of climatologically significant cloud-induced variations in the Earth's radiation budget, which is in line with the GMST growth in recent decades [11]. Another reason of AL growth may be the intensified wildfires in the NH boreal forests (due to droughts), especially in Canada and Alaska (see also [14,15]). In this region, the ST increase (Figure 1b,d) along with the precipitation reduction led to drought, which enhanced the wildfire severity [14].

Through changing ST, NDVI, AL and evapotranspiration, wildfire is the dominant driver of water, energy, and carbon cycles as well as vegetation dynamics in the North American boreal region [15,16]. The aforementioned region has the highest increase of AL that can reach up to 36% (Figure 1g). Likewise, Jin et al. [16] indicated that the increased severity of wildfires in the North American boreal forests during 2000–2009 enhanced AL by ~60%. Potter et al. [15] showed that climate change counteracts the cooling impact of postfire AL growth in the NH boreal forests. This justifies the increase of ST despite rising AL in the NH (Figure 1b,g).

Figure 1h compares the changes of ST, WS, and AL from 1961–1990 to 2000–2009 in the NH and SH. Both ST and AL increased in 86% of the NH and 9% of SH regions. Increase of ST in most regions on the NH leads to drought events, which intensify wildfires, ultimately rising AL. In addition, variations of ST are consistent with those of WS (i.e., both increase or decrease) in 53% and 51% of the areas of NH and SH, respectively. This happens because WS can affect ST through modifying evapotranspiration [17,18].

Regional studies are necessary to improve our understanding of the variations in ST, WS, and AL [4]. Some important regions for further investigations are the High Mountains of Asia, which demonstrate extreme values of ΔST (Figure 1b), ΔWS (Figure 1e), and ΔAL (Figure 1g) (see also [10,19]).

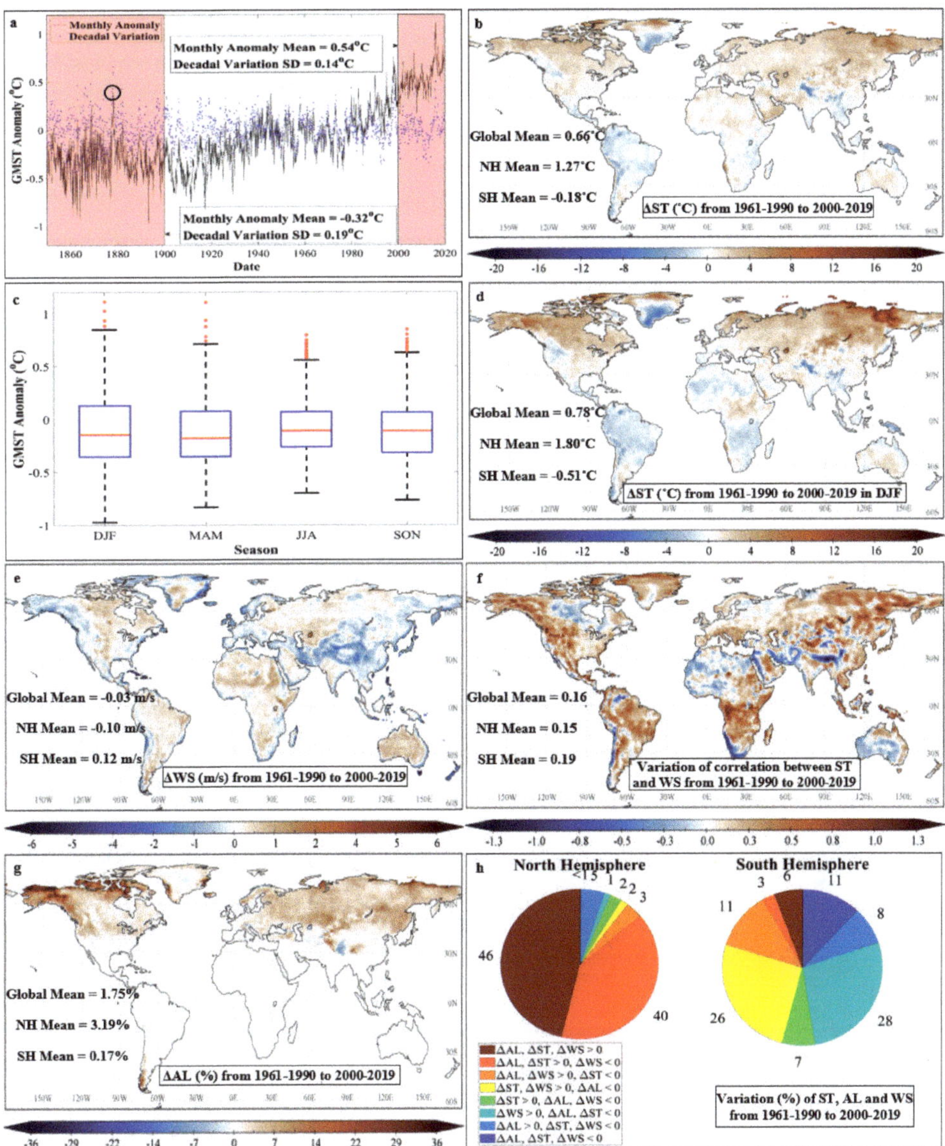

Figure 1. (**a**); Time series of monthly and decadal GMST anomalies. The 1850–1899 and 2000–2019 periods are shown by pink colors to compare the variations before the 20th century and after the 21st century, (**b**); variation of the GMST, (**c**); boxplots of the GMST anomalies. The horizontal line within the box indicates the median (50% percentile). The upper and lower edges of the box represent the 75% and 25% percentile, respectively. The upper and lower ends of the whiskers represent the maximum and minimum values. Outliers are observations beyond the end of the whiskers, (**d**); variation of the GMST in DJF, (**e**); variation of WS, (**f**); variation of the correlation between the GMST and WS, (**g**); variation of AL, (**h**); Hemispheric variation of the GMST, AL and WS. (**a**,**c**) are plotted based on the HadCRUT-version 4.6 product [20]. (**b**,**d**–**h**) are drawn based on the GLDAS (1° × 1° spatial resolution and monthly temporal resolution) dataset [21].

At the end, we would like to thank all the authors for their contributions to this Special Issue. We hope that the papers published in this Special Issue improve our knowledge in terms of application of climate data in hydrologic models.

Author Contributions: Conceptualization, M.V.; investigation, M.V., S.M.B., C.J.; writing—original draft preparation, M.V.; writing—review and editing, S.M.B., C.J., supervision, S.M.B., C.J. All authors have read and agreed to the published version of the manuscript.

Funding: This research received no external funding.

Institutional Review Board Statement: Not applicable.

Informed Consent Statement: Not applicable.

Data Availability Statement: The data that support the findings of this study are available from the corresponding authors upon reasonable request.

Conflicts of Interest: The authors declare no conflict of interest.

References

1. Hansen, J.; Sato, M.; Ruedy, R.; Lo, K.; Lea, D.W.; Medina-Elizade, M. Global temperature change. *Proc. Natl. Acad. Sci. USA* **2006**, *103*, 14288–14293. [CrossRef] [PubMed]
2. Foster, G.; Rahmstorf, S. Global temperature evolution 1979–2010. *Environ. Res. Lett.* **2011**, *6*, 044022. [CrossRef]
3. Hawkins, E.; Ortega, P.; Suckling, E.; Schurer, A.; Hegerl, G.; Jones, P.; Joshi, M.; Osborn, T.J.; Masson-Delmotte, V.; Mignot, J.; et al. Estimating changes in global temperature since the preindustrial period. *Bull. Am. Meteorol. Soc.* **2017**, *98*, 1841–1856. [CrossRef]
4. Huntingford, C.; Jones, P.D.; Livina, V.N.; Lenton, T.M.; Cox, P.M. No increase in global temperature variability despite changing regional patterns. *Nature* **2013**, *500*, 327–330. [CrossRef] [PubMed]
5. Nijsse, F.J.; Cox, P.M.; Huntingford, C.; Williamson, M.S. Decadal global temperature variability increases strongly with climate sensitivity. *Nat. Clim. Chang.* **2019**, *9*, 598–601. [CrossRef]
6. Wang, X.; Li, J.; Sun, C.; Liu, T. NAO and its relationship with the Northern Hemisphere mean surface temperature in CMIP5 simulations. *J. Geophys. Res. Atmos.* **2017**, *122*, 4202–4227. [CrossRef]
7. Fogt, R.L.; Marshall, G.J. The Southern Annular Mode: Variability, trends, and climate impacts across the Southern Hemisphere. *Wiley Interdiscip. Rev. Clim. Chang.* **2020**, *11*, e652. [CrossRef]
8. Falvey, M.; Garreaud, R.D. Regional cooling in a warming world: Recent temperature trends in the southeast Pacific and along the west coast of subtropical South America (1979–2006). *J. Geophys. Res Atmos.* **2009**, *114*, D04102. [CrossRef]
9. Vose, R.S.; Easterling, D.R.; Gleason, B. Maximum and minimum temperature trends for the globe: An update through 2004. *Geophysic. Res. Lett.* **2005**, *32*, L23822. [CrossRef]
10. Vautard, R.; Cattiaux, J.; Yiou, P.; Thépaut, J.N.; Ciais, P. Northern Hemisphere atmospheric stilling partly attributed to an increase in surface roughness. *Nat. Geosci.* **2010**, *3*, 756–761. [CrossRef]
11. Pallé, E.; Goode, P.R.; Montanes-Rodriguez, P.; Koonin, S.E. Changes in Earth's reflectance over the past two decades. *Science* **2004**, *304*, 1299–1301. [CrossRef] [PubMed]
12. Ghimire, B.; Williams, C.A.; Masek, J.; Gao, F.; Wang, Z.; Schaaf, C.; He, T. Global albedo change and radiative cooling from anthropogenic land cover change, 1700 to 2005 based on MODIS, land use harmonization, radiative kernels, and reanalysis. *Geophys. Res. Lett.* **2014**, *41*, 9087–9096. [CrossRef]
13. Wielicki, B.A.; Wong, T.; Loeb, N.; Minnis, P.; Priestley, K.; Kandel, R. Changes in Earth's albedo measured by satellite. *Science* **2005**, *308*, 825. [CrossRef] [PubMed]
14. Xiao, J.; Zhuang, Q. Drought effects on large fire activity in Canadian and Alaskan forests. *Environ. Res. Lett.* **2007**, *2*, 044003. [CrossRef]
15. Potter, S.; Solvik, K.; Erb, A.; Goetz, S.J.; Johnstone, J.F.; Mack, M.C.; Randerson, J.T.; Roman, M.O.; Schaaf, C.L.; Turetsky, M.R.; et al. Climate change decreases the cooling effect from postfire albedo in boreal North America. *Glob. Chang. Biol.* **2020**, *26*, 1592–1607. [CrossRef] [PubMed]
16. Jin, Y.; Randerson, J.T.; Goetz, S.J.; Beck, P.S.; Loranty, M.M.; Goulden, M.L. The influence of burn severity on postfire vegetation recovery and albedo change during early succession in North American boreal forests. *J. Geophys. Res. Biogeosci.* **2012**, *117*. [CrossRef]
17. Sheffield, J.; Wood, E.F.; Roderick, M.L. Little change in global drought over the past 60 years. *Nature* **2012**, *491*, 435–438. [CrossRef] [PubMed]
18. Valipour, M.; Bateni, S.M.; Gholami Sefidkouhi, M.A.; Raeini-Sarjaz, M.; Singh, V.P. Complexity of forces driving trend of reference evapotranspiration and signals of climate change. *Atmosphere* **2020**, *11*, 1081. [CrossRef]
19. Kraaijenbrink, P.D.; Bierkens MF, P.; Lutz, A.F.; Immerzeel, W.W. Impact of a global temperature rise of 1.5 degrees Celsius on Asia's glaciers. *Nature* **2017**, *549*, 257–260. [CrossRef]
20. HadCRUT. 2021. Available online: https://crudata.uea.ac.uk/cru/data/temperature/ (accessed on 10 May 2021).
21. GES DISC. 2021. Available online: https://disc.gsfc.nasa.gov/ (accessed on 10 May 2021).

Article

Traditional Water Governance Practices for Flood Mitigation in Ancient Sri Lanka

Vindya Hewawasam [1,*] and Kenichi Matsui [2]

[1] Graduate School of Life and Environmental Sciences, University of Tsukuba, 1-1-1 Tennodai, Tsukuba 305-8577, Japan

[2] Faculty of Life and Environmental Sciences, University of Tsukuba, 1-1-1 Tennodai, Tsukuba 305-8577, Japan; matsui.kenichi.gt@u.tsukuba.ac.jp

* Correspondence: vindyawh@yahoo.com

Abstract: The tank cascade system, which emerged as early as the fifth century BC in Sri Lanka's dry zone, has been portrayed as one of the oldest water management practices in the world. However, its important function as flood management has not yet been thoroughly examined. In this paper, we argue that the main principle behind the tank cascade system is not only to recycle and reuse water resources by taking advantage of natural landscapes but also to control floods. This paper examines the evolution of traditional water management and flood mitigation techniques that flourished in pre-colonial Sri Lanka. This historical examination also sheds light on recent policies that exhibited renewed interests in revitalizing some aspects of the tank cascade system in Sri Lanka's dry zone. This paper shows how ancient Sinhalese engineers and leaders incorporated traditional scientific and engineering knowledge into flood mitigation by engendering a series of innovations for land use planning, embankment designs, and water storage technologies. It also discusses how this system was governed by both kingdoms and local communities. Water management and flood control were among the highest priorities in urban planning and management. The paper thus discusses how, for centuries, local communities successfully sustained the tank cascade system through localized governance, which recent revitalized traditional water management projects often lack.

Keywords: tank cascade system; dry zone; water governance; flood control; traditional knowledge; community participation; Sri Lanka

Citation: Hewawasam, V.; Matsui, K. Traditional Water Governance Practices for Flood Mitigation in Ancient Sri Lanka. *Climate* 2022, 10, 69. https://doi.org/10.3390/cli10050069

Academic Editors: Mohammad Valipour and Sayed M. Bateni

Received: 8 April 2022
Accepted: 11 May 2022
Published: 13 May 2022

Publisher's Note: MDPI stays neutral with regard to jurisdictional claims in published maps and institutional affiliations.

1. Introduction

Water management is one of the fundamental requirements for the survival and prosperity of civilizations. Historically each civilization developed its unique water management practices that reflected the surrounding topography, climate, soil and utility purposes [1,2]. The traditional water management system that developed in the dry zone of Sri Lanka more than 2500 years ago is one of the oldest known water management systems in the world [3]. This ancient hydraulic civilization uniquely engineered storage dams [4] and water distribution systems [5,6].

According to the *Mahavamsa*, an earliest known chronicle that depicted ancient Sri Lanka, when Indian Prince Wijaya landed Sri Lanka from India in 543 BC, he observed irrigation practices among Indigenous Sinhalese people. At the port of Mannar, where he landed, the prince found many water tanks (or reservoirs) with cool water that replenished a great garden [5].

In the twenty-first century, this traditional water management is still in practice to some extent although much of it has been disrespected due to the introduction of Western water management systems under the colonial regime as well as in the age of the more contemporary international cooperation regime. However, somewhat reversing this trend, the Sri Lankan government has recently acknowledged the importance of historically practiced water management and reinstalled some in rural regions for flood protection and irrigation purposes.

Some scholars have reexamined Sri Lanka's traditional water management system partly for the purpose of enhancing modern day climate resilience actions. Some emphasized the implications for drought mitigation and rain water harvesting [7–11], agricultural developments [5,12,13], ecosystem management [2,14–17], and socio-economic development [4,18–20]. Literature shows that a similar tank system existed in semiarid southern India, but its main purpose was to provide water for paddy cultivation [21,22]. Qanat is another traditional irrigation water management practices that existed in semi-arid regions of Morocco, Spain, Syria, Iran, and Central and Eastern Asia. Ancient societies developed underground networks for the transportation of water [1].

Looking at this growing trend of studies, what is missing is a linkage between traditional water management and flood mitigation practices in Sri Lanka and elsewhere. Some researchers did mention about traditional flood mitigation functions in Sri Lanka [2,13,22–25], but the question remains as to the extent to which traditional water management practices were systemically arranged for improving or supplementing flood protection. In other words, we argue, flood mitigation has long been integral part of Sri Lanka's water management system.

This paper, therefore, seeks to understand Sri Lanka's traditional flood mitigation functions and technologies that evolved through time in its dry zone. This said, some may argue that this type of examination requires hydrological modeling or engineering investigation to truly understand the effectiveness of ancient flood mitigation infrastructure [26]. However, our main focus in this paper is rather to trace how past practices took shape in time, given urgent needs of local Sri Lankan farmers to mitigate flood risks. The IPCC report and other recent studies on climate change adaptation and disaster mitigation emphasized the importance of better understanding locally developed adaptation and mitigation practices as a way to enhance local disaster response capacity and participation [6,13,22,27–30].

In the following discussion, we first look at the development process and functions of the traditional system. Then, we examine water and flood management practices in ancient cities. Finally, we discuss the water governance system and its sustainability. For our examination, we used historical records, secondary sources, institutional reports, and audiovisual sources. In March 2019, the authors visited several ancient water management sites, including Sigiriya and Polonnaruwa to collect documentary and visual information. We also collected information in Colombo in the same year to find out what has already been known in the country about its ancient water management system.

2. Development of the Tank Cascade System in the Dry Zone of Sri Lanka

In Sri Lanka's dry zone annual mean precipitation is about 1750 mm whereas annual mean evaporation ranges from 1700 mm to 1900 mm [31]. About 80% of the annual rainfall occurs during the northeast monsoon season from November to February when flash floods often occur. Seasonal rivers and so-called *Villu* (wetland ecosystem in floodplains) are natural water bodies that emerge during these months. The earliest inhabitants were recorded in the lowland areas such as Anuradhapura and north central parts of Sri Lanka in the ninth century BC [32,33]. They lived along rivers and water bodies, collecting and storing water partly for drinking and irrigation purposes [2,6].

The earliest available information on hydraulic civilization of Sri Lanka dated back to the sixth to fifth century BC [4,5]. Archaeological studies show a network of tanks that were interconnected by streams and waterways. Today, this water network is commonly known as the tank cascade system [5,8,9,34–36]. The main hydrological principle behind the tank cascade system is to recycle and reuse water through a network of small to large scale tanks within a catchment. It also considers storing, transporting and distributing water for mitigating floods and droughts [2]. The International Union for Conservation of Nature in Sri Lanka identified four main functions of these ancient structures: (1) capturing rainwater to minimize floods; (2) storing rainwater; (3) recycling used water; and (4) mitigating drought impact [25,27]. Other than these, the cascade system sustained the local ecosystem as ancient engineers carefully used natural landscapes to enhance water storage [37].

The flood mitigation of the tank cascade system entailed engineering techniques of water and sedimentation flow control along with the protection of banks from erosion. Ancient Sinhalese people constructed granite structures and pillars. In order to protect the embankment from breaching and flooding, with improved technologies for metallurgy, iron was used to strengthen the structure [5]. The knowledge of iron metallurgy was introduced to Sri Lanka as early as the tenth century BC. The archaeological sites of Aligala, Sigiriya and Anuradhapura show evidence of iron smelting in the ninth century BC [38]. The construction of large tanks emerged in the fourth century BC. At the time of increasing water levels, tanks had spillways to safely release excess water from one tank to another. Small tanks were built in low plains between hills by connecting them with embankments [36]. For example, *Basawakkulama* tank, the earliest recorded large-scale tank was built in the Malwathu Oya Basin in about 430 BC. With over 3000 feet in length and 21 feet in height, the dam had storage capacity for cultivating 350 acres. A large number of small tanks were built in the same basin to avoid possible disasters from flooding [4].

Mahatantila et al. [39] identified three main components of tanks: (1) upper periphery, (2) bund/embankment and (3) tank body. However, in the following discussion, we add one more component, which is especially important for flood management; that is, the lower periphery of the tank where human settlements with paddies were located. Figure 1 shows how these components were typically laid out. Paddy cultivation was the main livelihood practiced by early inhabitants. The paddy cultivation of Sri Lanka dates back to the ninth to sixth centuries BC. During this period, ancient farmers domesticated cattle. Cattle were used for harrowing paddy fields [32,33,40].

Figure 1. Main components of the tank.

In the upper periphery of the tank running water in streams was filtered through forests with patchy water holes or bogs. Rain-fed farms were located here [2]. The ancient law prohibited felling trees as forests were important to manage water quantity and quality. Ancient people developed water holes (*godawala*) partly to prevent sediments from entering the tank [19,23]. Below these water holes, a water filtering area called *perahana* was created with water grasses like reeds [23,27].

Ancient Sinhalese protected the embankment from wind, heavy rain and waves by building stone liners on embankment walls [2,20]. The tank embankment was basically made of earth and granite rocks. Large-scale embankments with 30–40 feet deep reservoirs consisted of unique and intricate engineering innovations. The height of the embankment was carefully designed not to flood the upper stream area [14]. The embankment was installed with a sluice gate (*sorowwa*), valve pit (*bisokotuwa*), water level indicator (*diyakata pahana*), spillway (*pitawana*) and embankment protector (*ralapanawa*). The main purpose of sluice gate (*sorowwa*) was to regulate water release without flooding lower stream areas. The water level indicator helped decide when to release water. The valve pit or *bisokotuwa*, which was attached to the sluice gate on the bottom of the embankment, was basically a rectangular buffer room that was created to temporarily gather water from the lake through the sluice. When the level of gathered water in the room was raised above the sluice gate level but below the reservoir water level, water was released toward the lower periphery through the other gate(s). The location and arrangement of these water gates differed by region and embankment, showing engineering diversity [16,41]. The *bisokotuwa* is still in operation at Kalawewa, one of the largest tanks built in 477 AD during the period of Anuradhapura kingdom [2].

In the upper edge of tanks, a so-called tree belt (*gasgommana*) had a number of planted trees partly to protect the embankment. It also provided the habitat for fish and other aquatic species [2,23]. The trees became partly submerged in water during the heavy rain period. This tree belt also acted as a wind barrier and reduced the waves in the tank.

In the lower periphery of the tank, when water was released from a sluice gate through a valve pit it ran through an interceptor (*kattakaduwa*). The interceptor is a reserved land for the purpose of controlling soil erosion and water contamination. Villagers took drinking water below the interceptor. It also provided water to farms. The surrounding village was protected from water inundation with a hamlet buffer area (*thisbambe*), shrub land (*landa*) and drainage (*kivul ela*). The trees that have high heavy metal and salt absorption capacity with a strong root system were planted [2,23]. Being in the high elevated areas near the interceptor, villagers could observe flooding or damages to the embankment. The hamlet was surrounded by the hamlet buffer area that was used for common perennial cultivation (e.g., mango, coconut) and resting places for buffaloes [20,23]. Paddy fields below the interceptor functioned as wetlands during heavy rains to keep temporary flood water. When water is not enough for the whole paddy lands, all farmers cultivated equally (*Bethma cultivation*), limiting the paddy area to be irrigated [13,15]. The villagers used the shrub land for home gardening, such as chena cultivation. The excess water of the paddy fields flowed to the drainage area that was used for common village purposes to absorb salt and other contaminations [23,27]. Through the drainage and other natural streams, water reached the next tank.

3. Flood and Water Management Techniques in Ancient Kingdoms in Sri Lanka

In planning cities, villages and monastery complexes, ancient Sinhalese engineers carefully considered water sources, water uses and landscapes [42]. Flood control is one of the main requirements of city planning. Figure 2 shows the locations of the ancient kingdoms, main tanks and rivers. The historical records show that water was used not only for drinking and irrigation purposes but also for public bath and recreational activities. Traditional knowledge on rainfall patterns, land use planning and landscape helped ancient people maximize the use of water resources [17].

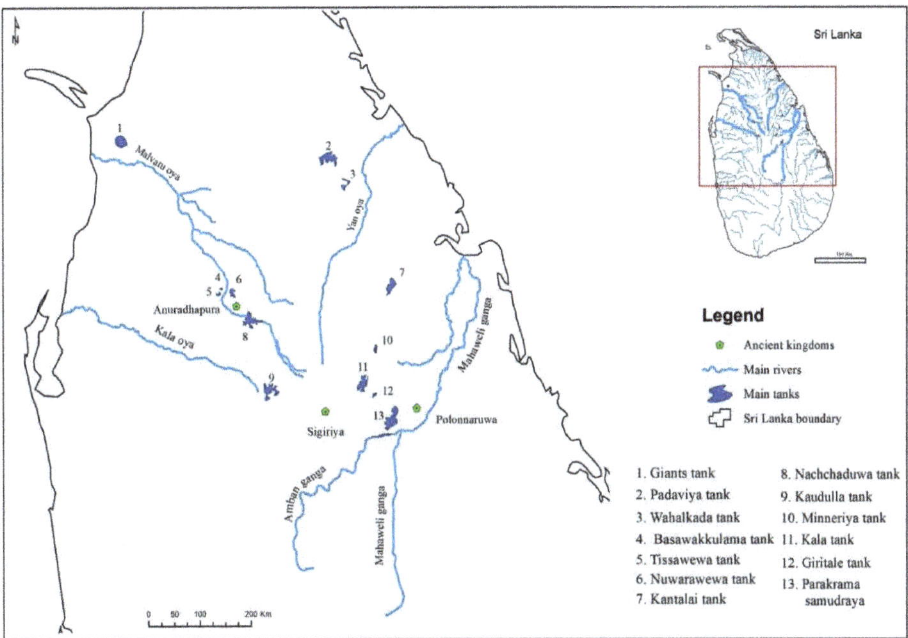

Figure 2. The locations of the ancient kingdoms, main tanks and rivers.

For example, in the early fifth century BC, the city of Anuradhapura adopted the ancient Indian "Nandyavarta plan" for water resource management [42]. It laid out the city in a circular shape. The central area was surrounded by walls with four gates that faced each cardinal direction. Anuradhapura was established along the Malwathu Oya (or river), the main river of the area. Anuradhapura's engineers constructed five large tanks around the city in about 437 BC. In order to protect the city from floods and droughts, they also constructed many small tanks in the same valley. Between the river and the tanks, three green parks were established mainly for recreational purposes [42]. These parks acted as water retention facilities during floods. Villages and king's palace were located below *Basawakkulama* tank. Its L-shaped embarkment was designed to take advantage of the surrounding landscape. It supplied both drinking and irrigation water. Bathing also became important part of Anuradhapura residents' lifestyle [42].

Sigiriya fortress, the present-day UNESCO World Heritage site, is another example to show a complex outlook of water management in an ancient city. The annual rainfall is the only possible water source here [8,17]. The fortress, which was built on the top of a gigantic rock as well as its surrounding areas, was built by King Kassapa in the fifth century AD (477–495). Human settlements began in this area as early as the third to the first century BC [43–45]. These early settlements were basically for monks who lived in caves of Sigiriya Rock. The rock walls just above caves were carved out like a gutter to keep out rainwater from flooding dwelling areas [43,45]. The cave entrances were then plastered for further protection. Archeologists identified about thirty such locations [43].

Later, King Kassapa developed an urban complex here [43,45]. Residents took water from the Sigiriya Oya and stored in a tank near Sigiriya Rock. Engineers at the time built storage tanks, cisterns, water-courses, underground and surface drainage to managed water in the city. All storage ponds and bathing pools were paved with marbles and pebbles to enhance water retention. In addition, natural depression areas were used to collect rainwater. The city was designed to control flow velocity, runoff discharge, and flow distance. For example, non-structural depression areas and drainage patterns were used to direct rainwater to the structural ponds located in lower elevations. During the process, water was filtered and velocity was reduced to control soil erosion [17]. Sigiriya also had water gardens and fountains [17,45]. The pools in the water gardens and Sigiriya tank were interconnected through underground drainage. This helped fill the pools automatically [8,42]. During the rain, the water garden can function as a water storage facility. In maintaining pools for bathing purposes, water was supplied from storage tanks, and the used water was released to moats through a separate drainage. The fountains were connected to special underground channels [42]. The moats were located in the lowest area of the land and excess water flows into the moats by reducing floods (Figure 3).

Figure 3. A water storage tank in the Sigiriya complex, Photo courtesy: Kenichi Matsui.

King's palace and the rock garden were located on the three-acre rock summit area about 360 m above the sea level [43]. Roofs were designed to collect and transport rainwater to the main water storage area. A drainage outlet was constructed to dispose excess water and prevent flooding. The surface area was terraced with the western side as the highest point [42].

After the demise of Sigiriya city, the Anuradhapura kingdom was reestablished in the 5th century AD about 80 km northwest from the rock. In its monastery site called Abayagiri, twin ponds were created in a low-lying area partly for monks to bathe. It is considered one of the best hydrologic engineering marvels of ancient Sri Lanka [46]. Underground pipelines were established to connect the ponds to Tissawewa, Basawakkulama, and Nuwarawewa tanks around the city by drawing water from the Malwathu Oya [42]. These pipelines ran through a number of small sediment/debris control tanks [47]. An enclosing wall was built around the ponds to control the possible spillage [46]. Also, wastewater outflows ran through wetlands for purification. Then the water was released back to the same river [8]. Each component of the ponds was carefully designed to protect the monastery complex from flooding.

Wastewater is a significant threat to health, particularly during flood events [28,48,49]. Ancient Sinhalese developed and practiced wastewater management. In Anuradhapura and Polonnaruwa different types of lavatories were developed [50]. Here urinals were collected in pits through terracotta pipes. Sands, lime powder and charcoals were used to purify wastewater [8,50]. In some places, separate septic tanks were used to store wastewater. These places were kept some distance from residential areas in order to avoid wastewater spillage during floods.

In the reign of King Parakramabahu (1153–1183 AD), water management technologies were further refined [5]. His engineers built several large-scale tanks and irrigation systems [23]. They constructed more than 163 major tanks, 2376 minor tanks, 165 anicuts and 3910 diversion channels [51]. The capital was located in present-day Polonnaruwa city, about 50 km east from Sigiriya. King's palace was located very close to the tank called *Parakyama Samudraya* (Figure 4) that had nine-mile-long embankment. In order to protect the palace from floods, huge brick walls were constructed along the tank side of the palace. There were several non-structural natural drainage facilities inside the walls to temporarily store water. The ground was also covered with grass to reduce soil erosion and trap debris. A few sluice gates were installed along the brick walls. A drainage canal was installed in the other side of the palace. *Kumara Pokuna* near King Parakramabahu's palace was one of bathing ponds that might have functioned as flood control structure (Figure 5). Similar to the Anuradhapura pipeline system it was connected with several drainages to purify water [17]. Even today this system is functioning well.

Figure 4. View of *Parakrama Samudraya* Photo courtesy: Kenichi Matsui.

Figure 5. *Kumara Pokuna* in Polonnaruwa Photo courtesy: Kenichi Matsui.

4. Water Governance in Ancient Sri Lanka

The ancient chronicles of *Mahavamsa*, *Dipavamsa* and *Culavamsa* as well as remaining cave, rock and slab inscriptions show evidence of Sri Lanka's water governance from the fourth century BC to the thirteenth century AD [3]. Some of these documents tell us how early Anuradhapura kingdom governance practices emerged with professionals and water ownership. The government imposed an income tax and other rules on using water [52]. These rules basically relied on community participation and involvement [10,13,17]. Later, the governance system gradually changed from a community-based system to a centralized one although the small-scale tank cascade system remained under community management [3].

From the fourth century BC to third century AD, water rights in general were held by individuals, kings, elites, local chiefs and families [3]. Kings and elites could grant water rights to monks mainly in the period between the second century BC and eighth century AD. Buddhist temples then administered water allocations. Until the second century BC, the management of tanks, including flood control and maintenance, were mainly undertaken at the village level, which mainly consisted of farmers [3,13]. Works for repair, desiltation and cleaning of the tank during the dry season were shared among farmers proportionately to the land ownership. Each farmer provided his or her service free on certain days [20]. The community also planted trees to strengthen the stability of the tank embankment and the interceptor (*Kattakaduwa*). In times of water shortage, the *Bethma* rules made farmers share water for paddy cultivation. This set of rules are still practiced today among some farmers during water shortage [13]. Here the village head or prominent leader decide the area for cultivation each year based on water availability in the tank. Farmers then received equity-based water allocation based on land ownership [13,20,40,53].

In the second century BC, localized water governance was gradually replaced with a centralized system. Different professions emerged as a result, such as flow operators (*Naguli*), canal officials, and proprietors of ferries (*Parumaka Thota Bojhaka*), and proprietor of tanks (*Parumaka vapihamika*). This institutionalized governance made it possible to sustain the food supply of a large population [3]. In the ninth century AD, the last phase of kingdom of Anuradhapura further institutionalized water governance by establishing specialized committees to maintain large-scale tanks [3].

Water rites, rituals, and customs played an important role in ancient water governance. For example, the king granted water rights through the "ceremony of golden vase", in which water was poured from a golden vase into farmers' hands [3]. The king also participated in festivals that sent a signal to commence ploughing, sowing, and harvesting. *Pen Pidima* ceremony offered fresh water of a tank to Buddha statues to pray for fertility [18].

5. Abandonment of Dry Zone Water Governance to Contemporary Water Governance

By the mid-13th century AD with the collapse of the Polonnaruwa kingdom, the centralized large-scale tank cascade governance was largely abandoned in many parts of the dry zone [3,4,31,54]. Although community-driven small-scale cascade systems remained in practice with varying degrees until the end of the 15th century [31]. European colonization under the Portuguese, Dutch, and British from the 16th through 18th centuries systemically and gradually disempowered traditional local authorities for water governance [4].

In 1832, about 17 years after the British established its colonial government in Kandy, it abolished the Kandyan *rajakariya* system, which imposed compulsory labor for public works, claiming that it resembled a form of slavery [3,31]. At the time many villages governed local affairs through *Gansabhawa*, a council composed of representatives of villagers. This council depended largely on the availability of village labor under the *rajakariya* system. As the British colonial regime further tightened restrictions on it, canals and other traditional water management works gradually fell into serious decay [54]. This change led to the deterioration of the community tank cascade system in many parts of the dry zone [3,31].

The British then empowered the Temple Land Commission (1856) and the Service Tenure Commission (1870) to govern tanks and surrounding villages where village heads used to control local affairs [3]. From 1870 to 1897, the colonial government repaired and restored a large number of village tanks, including *Kalawewa*, one of the largest tanks built in 477AD and Giant's Canal (*Yoda ela*), which was built in 459AD. Village communities provided their labor in voluntary basis for these tasks [54]. In 1900, as more educated Sri Lankan elites had been brought into civil services by then, the colonial government hired many of them and established the Department of Irrigation, which took control over public works. This department remains as one of the oldest departments in Sri Lanka. One of its mandates was flood protection [4,55]. During its 50-year operation, it restored almost all large-scale tanks and anicuts [4].

Soon after independence in 1948, the government of Sri Lanka attempted to improve water management. For example, in 1952, it constructed the nation's largest reservoir called *Senanayaka Samudraya* [29]. As the Gal Oya basin often experienced floods and droughts [55], disaster mitigation was one of the main components of the project. However, the project ultimately failed to achieve its initial goals due largely to top-down decision making where community participation was not incorporated [29].

In the 1950s, the Sri Lankan government began to place more emphasis on regional tank cascade governance. The Paddy Lands Act of 1958, for example, authorized the Department of Agrarian Services to maintain all small-scale water works. This act placed village tank cascade systems directly under the Department of Agrarian Services, a central government authority. It led the restoration and rehabilitation of small-scale village tanks including canals and flood mitigation structures [31].

In 1977, the Sri Lankan government undertook a comprehensive basin development called the Mahaweli Accelerated Development Project in the Mahaweli River Basin and created the Mahaweli Authority [4]. Flood control in the lower Mahaweli River was one of the main objectives [56,57]. It constructed new Western-style water reservoirs such as Kotmale, Victoria, and Maduru Oya. Kotmale reservoir was specifically designed for flood control by making it possible to transfer flood water from Polgolla to the dry zone where restored ancient tanks are located (e.g., *Parakrama Samudraya, Minneriya, Kantale, Kaudulla, Kalawewa and Giritale*) [58]. After the construction of this reservoir was completed in the mid-1980s, the flood inundation risk in the downstream of the Mahaweli River was significantly reduced [58]. The Mahaweli Authority also reestablished ancient river connections among Sudu Ganga, Amban Ganga, and Dambulla Oya in central province [59].

In the 1980s, the government undertook projects to restore ancient water management by mobilizing local people. In 1981, the Gal Oya Left Bank Rehabilitation Project hired community labor and collected local knowledge about water management [29,60]. It created farming organizations to control local water use for domestic and agricultural purposes. It also funded channel maintenance by these organizations. As farmers in the dry zone are the ones to experience flood damage to their crops, these farmer organizations were expected to play active roles in flood management [29]. In 1982, for example, the Village Irrigation Rehabilitation Project and National Irrigation Rehabilitation Project aimed to repair and maintain minor tanks with community participation [37].

In 1991, the Agrarian Services Act induced the concept of joint water management between farmers' organizations and a government agency [31]. In order to promote participatory planning and management, stakeholders were engaged in *kanna* meetings (pre-seasonal meetings of farmers). Even today, these meetings are the most important decision-making bodies in operating local tanks. Their tasks include the joint maintenance of flood control bunds, sluices, and channels [31].

Post-independent tank rehabilitation programs were mainly for repair, maintenance and physical development of individual tanks rather than the whole network of the tank cascade system. This shortcoming resulted in tank sedimentation, water leakage, land degradation, biodiversity loss, and floods and soil erosion [37]. In the 1990s, the Shared Control of Natural Resources in Watersheds Project adopted an ecosystem approach through community-based participatory watershed management. It expressed its strong interests in environmental sustainability and productivity improvement [60]. The project was implemented in northcentral province and southern province. It introduced farming companies to work independently in watershed management [61]. The project took the novel approach to address watershed issues by creating mini-projects among local communities and NGOs. This provided a sense of ownership for the parties involved in the project [62]. For example, in the Nilwala watershed, deforestation in the upper stream often flooded the lower basin due to sedimentation. To prevent soil erosion and sedimentation, the project attempted to protect existing forests and rehabilitate degraded forests. At villages, agroforestry practices were encouraged [63].

6. Conclusions

This paper has examined how Sri Lanka's traditional flood and water management evolved in its dry zone. Our goal was not to suggest that all traditional forms of flood control and community participation were effective. However, to better understand how Sri Lanka's flood management practices exist today in a hybrid form, which has incorporated traditional, colonial and modern technologies and practices, we found it imperative to clarify historical changes in flood management practices. Without knowing this complexity, locally viable flood mitigation measures cannot be sustainable. Therefore, it is important to understand how, in Sri Lanka, societies developed unique flood control practices, including governance frameworks and laws to ensure safety from disasters and equitable access to water. The tank cascade system is the result of their traditional knowledge on watershed and disaster management.

To demonstrate some recorded practices, this paper looked at how the tank cascade system captured rainwater, stored it, minimized flood impacts, maintained public health and conserved/nourished biodiversity. It somewhat resembled modern-day integrated water resources management. The traditional knowledge regarding engineering and metallurgy evolved and resulted in large-scale embankments for flood control along with sluice gate (*Sorowwa*), valve pit (*Bisokotuwa*), water level indicator (*Diyakata pahana*), spillway (*Pitawana*), and embankment protector (*Ralapanawa*).

Anuradhapura, Sigiriya, and Polonnaruwa cities developed both structural and non-structural water infrastructures. Multiple-purpose structures were designed for flood mitigation, irrigation, purification, drinking, and recreation. For centuries, the tank cascade system was largely governed by the community. Experienced community leaders played a vital role in decision making. Community voluntary support for managing the commons is an important feature in water governance. Rights to water resources were shared among elite groups, monks, community and individuals. The development of water professionals, taxes and rules in managing water systems emerged when centralized water governance under kingdoms began to exert more control over water resources. Along with these institutional development, monks and people developed water rituals and customs that considerably influenced traditional water governance.

European colonial regimes, however, gradually eroded this intricate water governance practices. The abolition of the *rajakariya system* under the British rule led to the disuse of communal tanks as it became difficult to obtain local labor. After independence, the government of Sri Lanka showed its renewed interests in traditional water governance and undertook several large-scale water management projects, such as the Gal Oya Irrigation Project and Mahaweli Accelerated Development Project with renewed interests in locally viable traditional water management. In the 1990s, Sri Lanka's watershed restoration policy began to emphasize community participation and led to some positive results. More water governance projects are planned to take advantage of traditional systems and mobilize local participation although the overall impact of their effectiveness for flood control under escalating climate change conditions remain to be determined in the future.

Author Contributions: Conceptualization, V.H. and K.M.; methodology, V.H.; software, V.H.; validation, V.H. and K.M.; formal analysis, K.M.; investigation, V.H. and K.M.; resources, V.H. and K.M.;

data curation, V.H. and K.M.; writing—original draft preparation, V.H.; writing—review and editing, K.M.; visualization, V.H.; supervision, K.M.; project administration, K.M. All authors have read and agreed to the published version of the manuscript.

Funding: This research received no external funding.

Institutional Review Board Statement: Not applicable.

Informed Consent Statement: Not applicable.

Data Availability Statement: Not applicable.

Conflicts of Interest: The authors declare no conflict of interest.

References

1. Behailu, B.M.; Pietilä, P.E.; Katko, T.S. Indigenous Practices of Water Management for Sustainable Services: Case of Borana and Konso, Ethiopia. Special Issue-Traditional Wisdom. *SAGE Open* **2016**, *6*, 2158244016682292. [CrossRef]
2. Geekiyanage, N.; Pushpakumara, D.K.N.G. Ecology of ancient Tank Cascade Systems in island Sri Lanka. *J. Mar. Isl. Cult.* **2013**, *2*, 93–101. [CrossRef]
3. Abeywardana, N.; Bebermeier, W.; Schütt, B. Ancient Water Management and Governance in the Dry Zone of Sri Lanka Until Abandonment, and the Influence of Colonial Politics during Reclamation. *Water* **2018**, *10*, 1746. [CrossRef]
4. Irrigation Department of Sri Lanka. Hydraulic Civilization and Water Resources in Sri Lanka. In Proceedings of the Hydro ASIA Fifth International Conference on Water Resources and Hydro Power Development in Asia, Colombo, Sri Lanka, 11–13 March 2014.
5. Brohier, R.L. *Ancient Irrigation Works in Ceylon. Part II. Northern and North-Western Sections of the Island*; Superintendent of Surveys; Ceylon Government Press: Colombo, Sri Lanka, 1935.
6. Dharmasena, P.B. Indigenous Water Resources Management in Sri Lanka. In Proceedings of the Conference on Environment and Humanity, Colombo, Sri Lanka, 19–20 August 2012.
7. Dharmasena, P.B. Indigenous Irrigation Practices. National Science Foundation, Sri Lanka. *Vidurava* **2004**, *32*, 3–7.
8. Turpin, T. *Traditional Water Management in Sri Lanka & Its Relevance to the UK for Climate Change*; Winston Churchill Memorial Trust: Colombo, Sri Lanka, 2006.
9. Jayasena, H.A.H.; Chandrajith, R.; Gangadhara, K.R. Water Management in Ancient Tank Cascade Systems in Sri Lanka: Evidence for Systematic Tank Distribution. *J. Geol. Soc. Sri Lanka* **2011**, *14*, 29–34.
10. Gangadhara, K.R.; Jayasena, H.A.H. *Rainwater Harvest by Tank Cascades in Sri Lanka Was It a Technically Adapted Methodology by the Ancients?* Department of Geology, University of Peradeniya: Peradeniya, Sri Lanka, 2014.
11. Piyadigama, T.M. Indigenous Water Resource Management in Sri Lanka. 2017. Available online: https://www.researchgate.net/publication/330957196_Indigenous_Water_Resource_Management_in_Sri_Lanka (accessed on 14 December 2019).
12. Dharmasena, P.B. Sustainability of Small Tank Irrigation Systems in Sri Lanka at the 21st Century. In *Traversing No Man's Land*; Godage International Publishers (Pvt) Ltd.: Matara, Sri Lanka, 2009; pp. 233–252.
13. Marambe, B.; Pushpakumara, G.; Silva, P. Biodiversity and Agrobiodiversity in Sri Lanka: Village Tank Systems. In *The Biodiversity Observation Network in the Asia-Pacific Region: Toward Further Development of Monitoring. Ecological Research Monographs*; Nakano, S., Yahara, T., Nakashizuka, T., Eds.; Springer: Tokyo, Japan, 2012. [CrossRef]
14. Panabokke, C.R. *The Small Tank Cascade System of the Rajarata*; Mahaweli Authority of Sri Lanka: Colombo, Sri Lanka, 2000.
15. Perera, M.P.; Chandima, D. Ancient Watershed Management Strategies in the Dry Zone of Sri Lanka. In Proceedings of the 3rd International Conference on Water & Flood Management (ICWFM), Dhaka, Bangladesh, 8–11 January 2011.
16. Dharmasena, P.B. Water Resources Management for Agriculture in the Dry Zone of Sri Lanka. A Review of the Journal of Tropical Agriculture. 2012. Available online: https://www.academia.edu/14024262/Water_Resources_Management_for_Agriculture_in_the_Dry_Zone_of_Sri_Lanka_A_Review_of_the_Journal_of_Tropical_Agriculture (accessed on 20 December 2020).
17. Gunasena, C. Structural and Non-Structural Water Management Concepts and Community Participation in Managing Land and Water Resources in Ancient Sri Lanka. Acedemia. 2018. Available online: https://www.academia.edu/36990554/Structural_and_nonstructural_water_management_concepts_and_community_participation_in_managing_land_and_water_resources_in_ancient_Sri_Lanka (accessed on 20 December 2020).
18. Abeysinghe, A. Historical Evidence of Water Management in Ancient Sri Lanka. *Econ. Rev.* **1980**, *6*, 6–7.
19. Wijesekera, S. Importance of Cascade Systems (Ancient Irrigation Systems) in Sustainable Development of Rural Communities in the Dry Zone of Sri Lanka (a Review of the Previous Studies). University of Kelaniya, Sri Lanka. 2015. Available online: http://repository.kln.ac.lk/handle/123456789/17710 (accessed on 14 December 2019).
20. FAO. *A Proposal for Declaration as a GIAHS. The Cascaded Tank-Village System (CTVS) in the Dry Zone of Sri Lanka*; Ministry of Agriculture, Food and Agriculture Organization of the United Nations: Colombo, Sri Lanka, 2017.
21. Gunnell, Y.; Krishnamurthy, A. Past and Present Status of Runoff Harvesting Systems in Dryland Peninsular India: A Critical Review. *Ambio* **2003**, *32*, 320–324. [CrossRef]

22. Bebermeier, W.; Meister, J.; Withanachchi, C.R.; Middelhaufe, I.; Schütt, B. Tank Cascade Systems as a Sustainable Measure of Watershed Management in South Asia. *Water* **2017**, *9*, 231. [CrossRef]

23. Dharmasena, P.B. Evolution of Hydraulic Societies in the Ancient Anuradhapura Kingdom of Sri Lanka. In *Landscapes and Societies*; Martini, I.P., Chesworth, W., Eds.; Springer: Dordrecht, The Netherlands, 2010; pp. 341–352. [CrossRef]

24. Schütt, B.; Bebermeier, W.; Meister, J.; Withanachchi, C.R. Characterisation of the Rota Wewa tank cascade system in the vicinity of Anuradhapura, Sri Lanka. *Die Erde J. Geogr. Soc. Berl.* **2013**, *144*, 51–68.

25. IUCN. Ecological Restoration of Kapiriggama Cascade Tank System. 2016. Available online: https://www.youtube.com/watch?v=DapXzbHLykQ (accessed on 14 December 2019).

26. Pedro, P.d.S.; Tavares, A.O. 2015. Basin Flood Risk Management: A Territorial Data-Driven Approach to Support Decision-Making. *Water* **2015**, *7*, 480–502. [CrossRef]

27. IUCN. *Restoring Traditional Cascading Tank Systems for Enhanced Rural Livelihoods and Environmental Services in Sri Lanka*; Cascade Development Project Information Brief No. 1; IUCN: Colombo, Sri Lanka, 2015.

28. Hughes, J.; Cowper-Heays, K.; Olesson, E.; Bell, R.; Stroombergen, A. Impacts and implications of climate change on wastewater systems: A New Zealand perspective. *Clim. Risk Manag.* **2021**, *31*, 100262. [CrossRef]

29. Uphoff, N. *Learning from Gal Oya: Possibilities for Participatory Development and Post-Newtonian Social Science*; Intermediate Technology Publications Ltd.: Rugby, UK, 1996.

30. IPCC. *Climate Change 2014: Impacts, Adaptation, and Vulnerability. Part A: Global and Sectoral Aspects*; Working Group II Contribution to the Fifth Assessment Report of the Intergovernmental Panel on Climate Change (IPCC); Cambridge University Press: Cambridge, UK, 2014.

31. Panabokke, C.R.; Sakthivadivel, R.; Weerasinghe, A.D. *Evolution, Present Status and Issues Concerning Small Tank Systems in Sri Lanka*; International Water Management Institute: Colombo, Sri Lanka, 2002.

32. Kulatilake, S. The Peopling of Sri Lanka from Prehistoric to Historic Times: Biological and Archaeological Evidence. In *A Companion to South Asia in the Past*; Schug, G.R., Walimbe, S.R., Eds.; John Wiley & Sons, Inc.: Hoboken, NJ, USA, 2016; pp. 426–436.

33. Premathilake, R. Relationship of environmental Changes in central Sri Lanka to possible prehistoric land-use and climate change. *Palaeogeogr. Palaeoclimatol. Palaeoecol.* **2006**, *240*, 468–496. [CrossRef]

34. Piyadasa, R.U.K.; Weerasinghe, M.; Weerakoon, L.; Somadewa, R.; Vipulasena, M.P. Sustainable Ancient Water Management and Resilience to Climate Change in Sri Lanka. In Proceedings of the International Conference on Climate Change Impacts and Adaptation for Food and Environment Security, Los Baños, Philippines, 21–22 November 2012.

35. Dharmasena, P.B. Water Balance of a Tank Cascade System in The Dry Zone. In *54th Annual Session of SLAAS*; Part 1:123; Sri Lanka Association for the Advancement of Science: Colombo, Sri Lanka, 1998.

36. Madduma Bandara, C.M. Tank cascade systems in Sri Lanka: Some thoughts on their development implications. In *Summaries of Papers Presented at Irrigation Research Management Unit Seminar Series during 1994*; Haq, K.A., Wijayaratne, C.M., Samarasekera, B.M.S., Eds.; IIMI: Colombo, Sri Lanka, 1995.

37. Dharmasena, P.B. *Guidelines for Tank Cascade Development*; Concept Note Submitted to Plan Sri Lanka for Tank Cascade Development Programme; Department of Agriculture, Field Crops Research and Development Institute: Maha Iluppallama, Sri Lanka, 2005.

38. Solangaarachchi, R. History of Metalaugury and Ancient Iron Smelting. Postgraduate Institute of Archology. University of Kelaniya. *Vidurawa* **1997**, *19*, 30–40.

39. Mahatantila, K.; Chandrajith, R.; Jayasena, H.A.H.; Ranawana, K.B. Spatial and temporal changes of hydro geochemistry in ancient tank cascade systems in Sri Lanka: Evidence for a constructed wetland. *Water Environ.* **2008**, *22*, 17–24. [CrossRef]

40. Dharmasena, P.B. Water use in Ancient Era. *Vidurava* **2001**, *29*, 3–9.

41. Peiris, K.; Wijesinghe, S. Introduction to the Function of Bisokotuwa in Ancient Vewa. *Engineer* **2008**, *41*, 24–28. [CrossRef]

42. Udalamatta, S.S. The Use of Water and Hydraulics in the Landscape Design of Sigiriya. Master's Thesis, Department of Architecture, University of Moratuwa, Moratuwa, Sri Lanka, 2003.

43. Bopearachchi, O. *The Pleasure Gardens of Sigiriya-A New Approach*; Godage International Publishers: Colombo, Sri Lanka, 2006.

44. Mandawala, P.B. *Buffer Zones in World Cultural Heritage Sites Sri Lankan Examples in Urban and Natural Setting*; President ICOMOS Sri Lanka; International Council on Monuments and Sites: Colombo, Sri Lanka, 2006.

45. Dumas, D.K. Sigiriya: An Early Designed Landscape in Sri Lanka. *Orientations* **2018**, *49*, 112–121.

46. Karunachandra, T.D. Kuttam Pokuna (Twin Ponds). Acedemia. University of Moratuwa. Sri Lanka. 2018. Available online: https://www.academia.edu/42989540/KUTTAM_POKUNA_twin_ponds_?email_work_card=view-paper (accessed on 5 April 2022).

47. Allceylon. Kuttam Pokuna (Twin Ponds). All Ceylon Tourism and Directory. Updated 20 October 2017. Available online: https://www.allceylon.lk/place/95-Kuttam-Pokuna-Twin-Ponds (accessed on 12 December 2019).

48. Han, J.; He, S. Urban flooding events pose risks of virus spread during the novel coronavirus (COVID-19) pandemic. *Sci. Total Environ.* **2021**, *755*, 142491. [CrossRef]

49. EPA. *Flood Resilience: A Basic Guide for Water and Wastewater Utilities*; EPA 817-B-14-006; Environmental Protection Agency: Washington, DC, USA, 2014.

50. Siriweera, W.I. Sanitation and Health Care in Ancient Sri Lanka. *Sri Lanka J. Humanit.* **2004**, *XXIX & XXX*, 1–12.

51. Jayasundera, M. Ancient Sri Lankan Economic Systems. *Financ. Mark. Inst. Risks* **2017**, *1*, 24–27. [CrossRef]

52. Abeywardana, N.; Pitawala, H.M.T.G.A.; Schütt, B.; Bebermeier, W. Evolution of the dry zone water harvesting and management systems in Sri Lanka during the Anuradhapura Kingdom; a study based on ancient chronicles and lithic inscriptions. *Water Hist.* **2019**, *11*, 75–103. [CrossRef]
53. Dharmasena, P.B. Essential Components of Traditional Village Tank Systems. In Proceedings of the Seminar on Cascade Irrigation Systems for Rural Sustainability, Colombo, Sri Lanka, 9 December 2010.
54. Mills, A.L. *Ceylon Under British Rule 1795-1932: With an Account of the East India Company's Embassies to Kandy 1762–1795*; Frank Cass and Co. Ltd.: London, UK, 1933.
55. Government of Sri Lanka. *Flood Protection Ordinance. Flood Protection Ordinance No 4 of 1924*; Government of Sri Lanka: Colombo, Sri Lanka, 1924.
56. Mohamed, Z. Development of Integrated Water Resources Management Plan for Eastern Dry Zone in Sri Lanka: The Case of Gal Oya River Basin. 2019. Available online: https://www.pwri.go.jp/icharm/training/master/img/2019/synopses/08_Zuhail_synopsis.pdf (accessed on 14 December 2019).
57. Chandrathilaka, W.A. Water Resources Management in Mahaweli and Adjoining River Basins in Sri Lanka. Engineer in Charge. Kotmale Dam and Reservoir. Mahaweli Authority of Sri Lanka. 2013. Available online: http://k-learn.adb.org/system/files/materials/2013/12/201312-water-resources-management-mahaweli-and-adjoining-river-basins-sri-lanka.pdf (accessed on 20 December 2020).
58. Rathnayake, U.; Weerakoon, S.B.; Nandalal, K.D.W.; Rathnayake, U. Flood Modeling in the Mahaweli River Reach from Kotmale to Polgolla. In Proceedings of the International Conference on Mitigation of the Risk of Natural Hazards, Colombo, Sri Lanka, 25–26 March 2007.
59. Weeraratne, S.; Wimalawansa, S.J. A Major Irrigation Project (Accelerated Mahaweli Programme) and the Chronic Kidney Disease of Multifactorial Origin in Sri Lanka. *Int. J. Environ. Agric. Res.* **2015**, *1*, 16–27.
60. Somasiri, H.P.S. Participatory Management in Irrigation Development and Environmental Management in Sri Lanka. *J. Dev. Sustain. Agric.* **2008**, *3*, 55–62.
61. Wijayaratne, C.M. Shared Control of Natural Resources (SCOR): An Integrated Watershed Management Approach to Optimize Production and Protection. *Sri Lankan J. Agric. Econ.* **1994**, *4*, 60–97.
62. Wijayaratne, C.M.; Jayawardena, K. Shared Control of Natural Resources (SCOR): A Market Oriented Watershed Management Strategy. Executive Summary. Paper 73 of Session 10. CGSpace. 1998. A Repository of Agricultural Research Outputs. Available online: https://cgspace.cgiar.org/handle/10568/37999 (accessed on 19 December 2021).
63. Elkaduwa, W.K.B.; Sakthivadivel, R. *Use of Historical Data as a Decision Support Tool in Watershed Management: A Case Study of the Upper Nilwala Basin in Sri Lanka*; Research Report 26; International Water Management Institute: Colombo, Sri Lanka, 1998.

Article

Climate Change Impacts on Groundwater Recharge in Cold and Humid Climates: Controlling Processes and Thresholds

Emmanuel Dubois [1,2,3,*], **Marie Larocque** [1,2,3], **Sylvain Gagné** [1,2] and **Marco Braun** [4]

1   Département des Sciences de la Terre et de l'atmosphère, Université du Québec à Montréal, Pavillon Président-Kennedy, C.P. 8888, Succursale Centre-Ville, Montréal, QC H3C 3P8, Canada; larocque.marie@uqam.ca (M.L.); gagne.sylvain@uqam.ca (S.G.)

2   GEOTOP Research Center, Université du Québec à Montréal, Pavillon Président-Kennedy, C.P. 8888, Succursale Centre-Ville, Montréal, QC H3C 3P8, Canada

3   GRIL Research Center, Département de Sciences Biologiques, Université de Montréal, Campus MIL, C.P. 6128, Succ. Centre-Ville, Montréal, QC H3C 3J7, Canada

4   Ouranos, 550 Rue Sherbrooke Ouest, Tour Ouest, 19e étage, Montréal, QC H3A 1B9, Canada; Braun.Marco@ouranos.ca

*   Correspondence: dubois.emmanuel@courrier.uqam.ca

Abstract: Long-term changes in precipitation and temperature indirectly impact aquifers through groundwater recharge (GWR). Although estimates of future GWR are needed for water resource management, they are uncertain in cold and humid climates due to the wide range in possible future climatic conditions. This work aims to (1) simulate the impacts of climate change on regional GWR for a cold and humid climate and (2) identify precipitation and temperature changes leading to significant long-term changes in GWR. Spatially distributed GWR is simulated in a case study for the southern Province of Quebec (Canada, 36,000 km^2) using a water budget model. Climate scenarios from global climate models indicate warming temperatures and wetter conditions (RCP4.5 and RCP8.5; 1951–2100). The results show that annual precipitation increases of >+150 mm/yr or winter precipitation increases of >+25 mm will lead to significantly higher GWR. GWR is expected to decrease if the precipitation changes are lower than these thresholds. Significant GWR changes are produced only when the temperature change exceeds +2 °C. Temperature changes of >+4.5 °C limit the GWR increase to +30 mm/yr. This work provides useful insights into the regional assessment of future GWR in cold and humid climates, thus helping in planning decisions as climate change unfolds. The results are expected to be comparable to those in other regions with similar climates in post-glacial geological environments and future climate change conditions.

Keywords: groundwater recharge; climate change; thresholds; seasonality; spatiotemporal variations; regional-scale; long-term; HydroBudget model; cold and humid climates; Quebec (Canada)

Citation: Dubois, E.; Larocque, M.; Gagné, S.; Braun, M. Climate Change Impacts on Groundwater Recharge in Cold and Humid Climates: Controlling Processes and Thresholds. *Climate* **2022**, *10*, 6. https://doi.org/10.3390/cli10010006

Academic Editors: Mohammad Valipour, Sayed M. Bateni and Ying Ouyang

Received: 15 November 2021
Accepted: 4 January 2022
Published: 12 January 2022

Publisher's Note: MDPI stays neutral with regard to jurisdictional claims in published maps and institutional affiliations.

1. Introduction

Climate change is already impacting the water cycle everywhere around the world because of precipitation changes and warming temperatures [1]. In particular, it is impacting surface water and groundwater systems in cold and humid climates due to high rates of precipitation change and warming temperatures [2–5]. Because changes at the surface propagate to aquifers through groundwater recharge (GWR), they could have major impacts on groundwater use for human consumption, industry, and agriculture, as well as on groundwater-dependent ecosystems [6–10]. Although the impacts of climate change on groundwater are increasingly studied, the uncertainty associated with simulations of future climatic conditions remains high [9,11–14]. This is even more remarkable in cold and humid climates, where precipitation changes are uncertain (increase or decrease) and where seasonal snow coverage, which leads the annual hydrological cycle, is particularly sensitive to cold season temperatures [2]. A literature review of climate change impacts on groundwater systems in eastern Canada highlighted the wide variability of simulation results from 22 studies spanning the Canadian provinces of Ontario, Quebec, New Brunswick, Nova Scotia, and Prince Edward Island [15]. Using different hydrological, hydrogeological, or integrated models, different downscaling techniques, or different time horizons thus adds further uncertainty to the analysis [15–17], making it difficult to compare future outlooks. Nevertheless, simulation of future conditions remains an essential tool for long-term groundwater resource management in a climate change context.

In cold and humid regions, the geomorphology has largely been shaped by the latest glaciation cycle, groundwater levels are often close to the surface, and unconfined aquifers are generally fed through GWR and connected to superficial water bodies [18–21]. Groundwater recharge is constrained during winter by freezing temperatures that reduce the available liquid water and during the growing season by intensive evapotranspiration rates [4,19,22]. Overall, hydrological systems are highly responsive to changes in the water cycle (e.g., spatio-temporal distribution of precipitation, rainfall intensity, snow accumulation, timing of snowmelt), thus propagating climate changes to the regional hydrology [23]. The impact of climate change in cold and humid regions characterized by an important rise of temperature in the future (especially during the winter season) and by uncertain future precipitation conditions [5,24–26], has been widely studied [2,4,27–33]. The decrease in snow water storage is recognized as a leading cause of low summer stream flows [29,34–36]. As winter temperatures increase, snow cover decreases and winter GWR events become more frequent and are associated with increased winter flow rates in rivers, as evapotranspiration remains very low [4,22,30,37]. Spring peaks of flow rates or GWR become subdued as snow-dominated hydraulic regimes transition to rain-dominated regimes [4,11,15,29,33,38–41]. Although not yet well understood, these changes can be important for all groundwater users (humans, industries, ecosystems) and thus need to be studied and forecasted to support future water use policies.

Most climate change impact studies identify ranges of changes in hydrologic variables associated with the climate forcing of the climate scenarios used for simulation [13,27,30,39,41,42]. Reineke et al. [17] observed statistically significant changes in global-scale GWR for different warming levels (+1 to +3 °C). Similarly, Döll et al. [7] presented a global analysis of additional hydrologic hazard occurrence resulting from +1.5 and +2 °C warming using hydrological indicators, including GWR. However, a range of changes can be insufficient to properly adapt infrastructures and policies to future conditions, as climate change signals usually overlap between climate models and RCPs. To overcome this, Crosbie et al. [13] provided data for water management scenarios using a scale of probability that simulated how future GWR would change from the simulated historical GWR at the Australian scale. Kløve et al. [9] suggested the use of indicators to communicate climate simulation results and representative parameters for use in water resource planning. These indicators of future conditions can be derived from winter low flows [27], GWR volumes [43], or water table depths [44], and can be expressed, for example, as functions of Quaternary deposits [27]. Meanwhile, identification of the evolution of precipitation and temperature changes that would lead to noticeable and statistically significant changes in GWR over time has not yet been undertaken. This assessment of the sensitivity of GWR to changes in climate variables, without a specific distinction between different climate forcing scenarios, would facilitate inter-study comparisons and provide simple and accessible indicators of future conditions for water managers [11].

This work aims to provide new insights into these questions. The objectives are (1) to simulate the effect of climate change on potential GWR in cold and humid climates and post-glacial geological environments and (2) to identify controlling processes and thresholds of GWR changes. As a regional-scale case study, this work focuses on future GWR conditions for the southern region of the Province of Quebec, Canada (36,000 km^2), where the hydrology is dominated by long and cold winters. A spatially distributed water budget GWR model calibrated over the 1961–2017 period [22] was used with a set of 12 climate scenarios downscaled from global climate models (GCMs) using RCP4.5 (low emissions) and RCP8.5 (high emissions). The results were used to identify the controlling processes of GWR changes, as well as temperature and precipitation thresholds that lead to significant long-term changes emerging from future climate uncertainty.

2. Data and Methods

2.1. Study Area

A detailed description of the study area can be found in Dubois et al. [22] and is summarized hereafter. The study area is located in the province of Quebec (humid continental climate), between the St. Lawrence River and the Canada–USA border, and between the Quebec–Ontario border and Quebec City (35,800 km^2) (Figure 1). It is comprised of the watersheds of eight main tributaries of the St. Lawrence River (numbered W1 to W8 from west to east) (Table 1). Watersheds W1, W2, and W4 have 42%, 83%, and 15% of their total areas, respectively, located in the USA. The topography is flat, with low-elevation areas close to the St. Lawrence River and higher elevations in the Appalachian Mountains. The land cover includes agriculture (42%), forest (45%), wetlands (6%), urban uses (5%), and surface water (2%). The annual average temperature varies between 6.5 (W1, west) and 3.9 °C (W8, east), with the west–east cooling gradient also being notable during the cold months (December to March, T < 0 °C). The annual precipitation ranges between 952 (W1) and 1123 mm/yr (W4), corresponding to an average of 231 (W3) to 142 mm (W7) of vertical inflow (VI; available liquid water = sum of rainfall and snowmelt) during the cold months.

Figure 1. Location of the study area and the watersheds within.

Table 1. Watershed characteristics, climate, and potential groundwater recharge (GWR) for the 1961–2017 period (from 22).

	Area (km^2)	Median Elevation (m asl)	Annual		Cold Months		Pot. GWR				
			T (°C)	P (mm)	T (°C)	VI (mm)	mm/yr	Win.	Spr.	Sum.	Fall
W1. Châteaugay *	2219	51	6.5	952	−6.4	211	109	38%	46%	3%	14%
W2. Richelieu *	4414	40	6.3	1039	−6.7	223	119	36%	45%	4%	15%
W3. Yamaska	4792	80	5.9	1080	−7.1	231	139	35%	44%	4%	17%
W4. Saint-François *	9068	312	4.8	1123	−7.8	214	147	31%	42%	8%	19%
W5. Nicolet	3591	150	5.1	1076	−8.0	196	144	32%	43%	6%	19%
W6. Bécancour	3380	140	4.5	1103	−8.7	164	151	28%	44%	7%	21%
W7. Du Chêne	461	90	4.5	1092	−8.9	142	154	26%	46%	8%	20%
W8. Chaudière	7879	340	3.9	1092	−8.9	151	145	27%	42%	10%	21%

* Part of the watershed is located in the USA—the presented values are only for the Quebec part. Cold months = DJFM; Win. = DJF; Spr. = MAM; Sum. = JJA; Fall = SON; VI = vertical inflow (available liquid water, the sum of rainfall and snowmelt).

The study area includes two main geological units, the sedimentary basin of the St. Lawrence Platform and the metasedimentary Appalachian Mountains. The bedrock is unevenly covered with unconsolidated Quaternary sediments from the last glaciation–deglaciation cycle and is mainly composed of thin and coarse superficial materials deposited on the bedrock in the uphill areas, thick and mixed-grain size deposits in the valleys, and clay covering sandy materials close to the St. Lawrence River. Regional fractured bedrock aquifers flow from the Appalachians to the St. Lawrence River (south/southeast to north/northwest). The aquifers are moderately productive and are in unconfined conditions upstream and semi-confined to confined conditions in the valleys and in the St. Lawrence Lowlands [21]. Dubois et al. [22] estimated the average 1961–2017 regional potential GWR to be 139 mm/yr. They identified preferential recharge zones in the Appalachians, in forested areas, and over coarse deposits and outcropping bedrock. The potential GWR increases from west to east (Table 1). Warmer temperatures in the western watersheds (W1 to W3) are responsible for higher winter GWR rates (more VI) and lower summer and fall GWR rates (more actual evapotranspiration, AET) than in the eastern watersheds. The peak of the spring GWR, which is linked to snowmelt in April, dominates GWR in all the watersheds and corresponds to 44% of the annual GWR rates.

2.2. Simulating Groundwater Recharge with Hydrobudget

The HydroBudget model (HB) is a water budget GWR model developed to compute spatially distributed potential GWR on grid cells of regional-scale watersheds over long periods of time [45,46]. Using the spatially distributed daily temperature, daily total precipitation, and runoff curve number (RCN—a combination of pedology, land cover, and slopes), HB is driven by eight parameters that require calibration (Table 2) and aggregates the output at a monthly time-step (although daily input data are required). For each daily time-step, HB computes the snow accumulation and melt (two parameters, T_M and C_M), tests if the soil is frozen (two parameters, TT_F and F_T), and partitions the superficial runoff (based on the RCN and with two parameters, t_{API} and f_{runoff}) from the water that infiltrates into a conceptual soil reservoir (one parameter, sw_m). The AET corresponds to the minimum between the potential evapotranspiration calculated using the formula of Oudin et al. [47] and the available water in the soil reservoir. Part of the residual soil reservoir water is mobilized as potential GWR (one parameter, f_{inf}). Since HB does not simulate percolation out of the unsaturated zone, the potential GWR represents a maximum of GWR that could reach the saturated zone. For simplification, the simulated potential GWR will be hereon referred to as GWR.

Table 2. Description of the HydroBudget parameters and calibrated values from Dubois et al. [22].

	Parameter		Regionally Calibrated Value from Dubois et al. [22]
Snowmelt model	Melting temperature—T_M (°C)	Air temperature threshold for snowmelt	1.4
	Melting coefficient—C_M (mm/°C/d)	Melting rate of the snowpack	4.9
Freezing soil conditions	Threshold temperature for soil frost—TT_F (°C)	Air temperature threshold for soil frost	−15.9
	Freezing time—F_T (d)	Duration of air temperature threshold to freeze the soil	16.4
Runoff	Antecedent precipitation index time—t_{API} (d)	Time constant to consider the soil in dry or wet conditions based on previous precipitation event	3.9
	Runoff factor—f_{runoff} (-)	Correction factor for runoff computed with the RCN method for the partitioning between runoff and infiltration into the soil reservoir	0.60
Lumped soil reservoir	Maximum soil water content—sw_m (mm)	Soil reservoir storage capacity, maximum height of water stored in a 1 m soil profile	385
	Infiltration factor—f_{inf} (-)	Fraction of soil water that produces deep percolation at each daily time step	0.06

Assuming that surface watersheds match hydrogeological watersheds and that rivers drain unconfined aquifers, Dubois et al. [22] calibrated HB by comparing the sum of the simulated superficial runoff and simulated GWR to the measured river flow and the simulated GWR to baseflow estimates (regressive filter on river flow time series). A multi-objective automatic calibration procedure was used with time series of river flow rates from 51 gauging stations over the 1961–2017 period and showed that the simulated variables had a small uncertainty ($\leq$10 mm/yr). Therefore, this regionally calibrated model was considered suitable to be used to simulate future GWR over the study area.

2.3. Climate Scenarios

A subset of 12 climate scenarios was derived from an ensemble of 54 climate simulations provided by 29 global climate models (GCMs) from the Coupled Model Intercomparison Project—Phase 5 (CMIP5) driven by RCP4.5 and RCP8.5 future greenhouse gas concentrations. The 12-member ensemble (Table 3) was created using the k-means clustering method proposed by Casajus et al. [48]. The climate scenario clustering process was constrained by ten criteria: the change in annual mean temperature and precipitation between the 1981–2010 period and the 2041–2070 period (two variables) and the changes in seasonal mean temperature and precipitation between the same periods (eight variables). Nine out of the 12 clusters comprised climate scenarios based on both RCPs. The algorithm selects the climate scenario closest to the cluster centroid as the candidate (not considering their RCP), as it best represents the average condition of future precipitation and temperature of the cluster. However, CanESM2 (CE2) was hand-picked from its respective cluster to include the Canadian GCM. The subset captures approximately 85% of the initial variance of the ensemble of 54 climate simulations. Casajus et al. [48] showed that this method retains a good representativity of the uncertainty linked to the climate scenarios between an ensemble of 27 climate scenarios and its subset of six.

Table 3. Selected climate scenarios.

Name.	Model Source	Code	RCP
ACCESS1-0_rcp45_r1i1p1	Commonwealth Scientific and Industrial Research Organization (CSIRO), Australia and Bureau of Meteorology (BOM), Australia	A10	4.5
ACCESS1-3_rcp85_r1i1p1	Commonwealth Scientific and Industrial Research Organization (CSIRO), Australia and Bureau of Meteorology (BOM), Australia	A13	8.5
bcc-csm1-1-m_rcp45_r1i1p1	Beijing Climate Center, China Meteorological Administration, China	B1M	4.5
BNU-ESM_rcp85_r1i1p1	College of Global Change and Earth System Science, Beijing Normal University (BNU), China	BNU	8.5
CanESM2_rcp45_r1i1p1	Canadian Center for Climate Modelling and Analysis (CCCma), Canada	CE2	4.5
CMCC-CMS_rcp45_r1i1p1	Centro Euro-Mediterraneo sui Cambiamenti Climatici Climate Model, Italy	CMS	4.5
GFDL-CM3_rcp45_r1i1p1	Geophysical Fluid Dynamics Laboratory (GFDL), USA	GF3	4.5
GISS-E2-R_rcp45_r6i1p3	National Aeronautics and Space Administration (NASA)/Goddard Institute for Space Studies (GISS), USA	GIR	4.5
inmcm4_rcp45_r1i1p1	Institute for Numerical Mathematics (INM), Russia	INM	4.5
MIROC-ESM _rcp45_r1i1p1	Japan Agency for Marine-Earth Science and Technology, Atmosphere and Ocean Research Institute (The University of Tokyo), National Institute for Environmental Studies, Japan	MIC	8.5
MIROC-ESM-CHEM_rcp85_r1i1p1	Japan Agency for Marine-Earth Science and Technology, Atmosphere and Ocean Research Institute (The University of Tokyo), National Institute for Environmental Studies, Japan	MIE	4.5
MRI-ESM1_rcp85_r1i1p1	Meteorological Research Institute, Japan	MRE	8.5

The 12 selected simulations were bias-corrected to a 1981–2010 reference dataset (Natural Resources Canada gridded observation database) [49,50] and downscaled to the reference 10 km × 10 km resolution using the quantile mapping approach by Mpelasoka and Chiew [51]. With these scenarios, changes in mean annual temperature and annual precipitation between the 1981–2010 and 2041–2070 periods (ΔT and ΔP, respectively) covered most of the combinations of ΔT and ΔP found in the ensemble of 54 climate scenarios (Figure 2a). ΔT ranged between +0.9 (INM, RCP4.5) and +5.0 °C (MIC, RCP8.5) and ΔP ranged between +5 (B1M, RCP4.5) and +200 mm (A13, RCP8.5). The change in mean temperature during the cold months (December to November; ΔT_{CM}) was between +1.1 (INM, RCP4.5) and +6.0 °C (MIC, RCP8.5) (Figure 2b). The change in precipitation during the cold months (ΔP_{CM}) was between +17 (MIE, RCP4.5) and +100 mm (GF3, RCP4.5). The warming temperature during the cold months led to ΔVI_{CM} between +33 (INM, RCP4.5) and +215 mm (MIC, RCP8.5) (Figure 2c).

2.4. Period Comparisons and Significant Changes

The simulation period was divided into four 30-year periods: 1981–2010, the reference period, also used as the baseline for the bias correction of the climate scenarios, and three future periods, 2011–2040, 2041–2070, and 2071–2100. The same periods were used to present the results of the 96 GWR scenarios (12 scenarios for 8 watersheds). Each 30-year period was compared to the previous one and to the 1981–2010 reference period to observe the simulated range in future GWR and to identify future GWR changes.

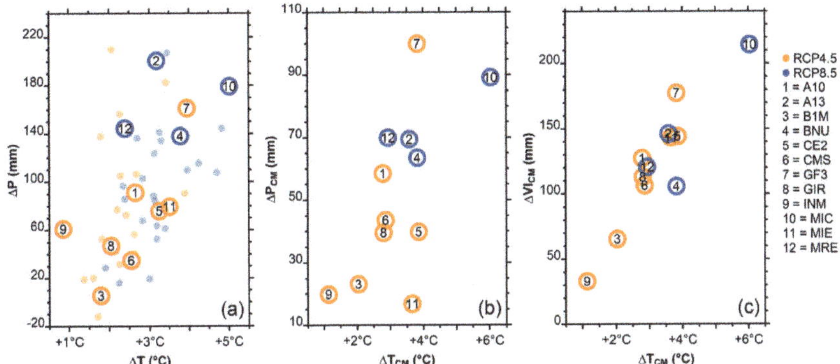

Figure 2. Changes (Δ) between the reference period (1981–2010) and the 2041–2070 horizon for the 12 selected climate scenarios in the study area in (**a**) annual precipitation (ΔP) as a function of mean annual temperature (ΔT) within the ensemble of 54 climate scenarios (27 RCP4.5 and 27 RCP8.5), (**b**) cold month (December to March) precipitation (ΔP_{CM}) as a function of mean cold month temperature (ΔT_{CM}), and (**c**) vertical inflow during the cold months (sum of liquid precipitation and snowmelt; ΔVI_{CM}) as a function of mean cold month temperature (ΔT_{CM}).

Changes in precipitation, temperature, and GWR were determined to be statistically significant based on the Tukey test ($p < 0.05$), comparing the results of each 30-year period and the previous one or between the future periods and the reference period. The sample size in each group varied between 30 values, when monthly or annual variables for each watershed and each scenario were compared, and 360 values for the monthly or annual variables for each watershed (or grid cells) when all the scenarios were compared.

3. Results

3.1. Climate Changes for the 1981–2100 Period

The average evolution of annual precipitation and mean temperature from the 12 scenarios for each watershed and for each 30-year period shows a constant increase throughout the century (Figure 3). All increases between each 30-year period and the previous one (30 years and 12 scenarios corresponding to 360 values for each period) were statistically significant. The range of precipitation and temperature changes, represented by the difference between the minimum and the maximum of the 12 scenarios for each year, increased remarkably from the 1981–2010 period to the 2071–2100 period (Figure 3).

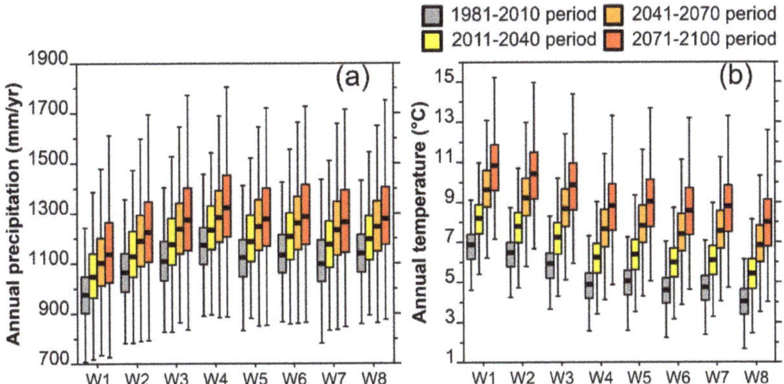

Figure 3. Thirty-year period change of mean annual precipitation (**a**) and annual temperature (**b**) of the 12 climate scenarios for the eight watersheds (W1 to W8).

3.2. Groundwater Recharge for the 1981–2100 Period

The previously calibrated HB model [22] was used to simulate future GWR under the 12 climate scenarios for the entire 1951–2100 period with a monthly time-step and a 500 m × 500 m spatial resolution. Although simulations were performed for the 1951–2100 period, GWR changes were only compared between three future periods (2011–2040, 2041–2070, 2071–2100) and the previous period (including the 1981–2010 reference period). The 30-year moving averages for all GWR scenarios ranged between 126 (W1) and 183 mm/yr (W6) (Figure 4). The ensemble mean GWR increased between +5 (W8) and +17 mm/yr (W1) from the 1981–2010 period to the 2071–2100 period, with maximum increases of +5 mm/yr between two consecutive 30-year periods (Table 4). Six climate scenarios produced GWR rates higher than the ensemble mean (A13, BNU, CMS, GIR, MIC, MRE, RCP8.5, and RCP4.5). The other six climate scenarios (A10, B1M, CE2, GF3, INM, MIE, and RCP4.5) produced GWR rates lower than the ensemble mean. Climate scenarios based on RCP8.5 produced higher GWR rates (although not always the highest).

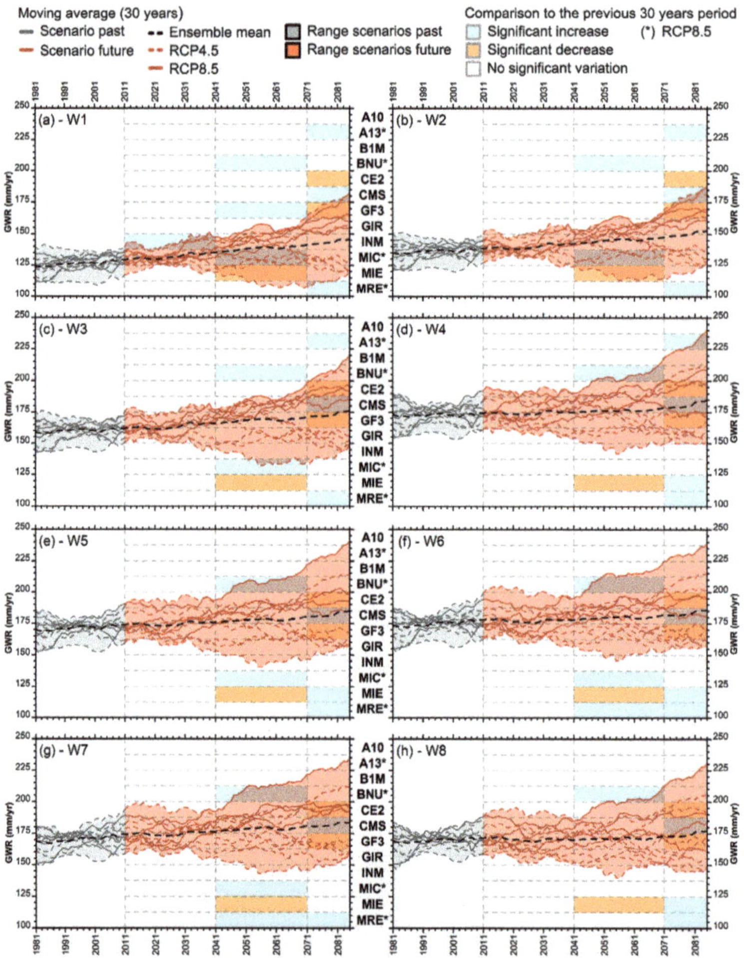

Figure 4. Thirty-year moving average of groundwater recharge (GWR) simulated with the 12 climate scenarios and significant changes (Tukey test, $p < 0.05$) between subsequent 30-year periods (2011–2040, 2041–2070, and 2071–2100) for (**a**) W1 to (**h**) W8.

Table 4. Thirty-year evolution of mean groundwater recharge (mm/yr), range of the ensemble changes (in brackets), and evolution of the cold month groundwater recharge from December to March (T < 0 °C) and that from May to November (T > 0 °C) (mm) of the 12 climate scenarios for the eight watersheds (W1 to W8).

	W1		W2		W3		W4		W5		W6		W7		W8	
	CM	WM	CM	WM	CM	WM	CM	WM	CM	WM	CM	WM	CM	WM	CM	WM
1981–	126 (26)		136 (24)		160 (26)		173 (28)		171 (26)		175 (28)		170 (28)		170 (24)	
2010	74	31	74	39	80	53	75	71	77	66	71	74	66	72	66	77
2011–	132 (22)		140 (22)		164 (26)		175 (34)		175 (34)		177 (38)		175 (36)		171 (36)	
2040	82	27	82	33	89	45	85	61	86	57	81	64	77	64	75	67
2041–	138 (42)		145 (42)		168 (48)		176 (54)		178 (60)		180 (60)		178 (60)		171 (50)	
2070	93	25	93	30	102	41	96	54	99	51	94	57	90	58	85	58
2071–	143 (60)		150 (60)		173 (68)		181 (76)		182 (78)		183 (76)		182 (76)		175 (74)	
2100	101	22	102	27	112	36	107	48	110	45	105	50	102	50	97	51

CM = sum of groundwater recharge for the "cold months", from December to March (T < 0 °C); WM = sum of groundwater recharge for the "warm months", from May to November (T > 0 °C).

The range of changes in the GWR scenarios was smallest for the 1981–2010 period (2011–2040 for W1 and W2), with values between 22 (W1 and W2) and 28 mm/yr (W4, W6, and W7) (Table 4). It increased markedly in the 2041–2070 period, with values between 42 (W1 and W2) and 60 mm/yr (W5, W6, and W7). It increased even more during the 2071–2100 period, reaching values of between 60 (W1 and W2) and 78 mm/yr (W5). This larger range of the results was due to the increasing range in precipitation changes between the scenarios in the second half of the 21st century (Figure 3).

The climate changes associated with each significant ΔGWR between 30-year periods (not using the 30-year moving average) are reported in Table 5. Although there was a general increase in temperature between 1981 and 2100, a relatively small number of significant inter-period ΔGWR were observed (Figure 4). This could be due to the combined effect of increased evapotranspiration triggered by higher temperature and increased precipitation. As such, the direction of the ΔGWR change was not directly linked to the change in precipitation (ΔP). For example, ΔGWR > 0 (increase) was associated with ΔP < 0 (CMS for the 2071–2100 period, compared to 2041–2070 for W3 and W8), while ΔGWR < 0 (decrease) was simulated with ΔP > 0 (CE2 and GF3 for the 2071–2100 period compared to 2041–2070 for W3 and W8; MIE for the 2041–2070 period compared to 2011–2040 for W3, W6, and W8). An average temperature change (ΔT) of +1.2 °C was associated with ΔGWR < 0 (between +0.7 and +2.3 °C), while an average ΔT of +1.8 °C was associated with ΔGWR > 0 (between +0.2 and +2.8 °C). The four climate scenarios based on RCP8.5, representing mainly very humid future conditions, produced statistically significant ΔGWR > 0 for one of the last two future periods, except for A13 in the eastern watersheds (W5 to W8; Figure 4e–h). The climate scenarios based on RCP4.5, representing both moderately and very humid future conditions, produced both ΔGWR < 0 and ΔGWR > 0 for different periods. In addition, the changes between the 1981–2010 and the 2011–2040 periods were not statistically significant (only one significant change in W1 for one scenario; Figure 4a).

Overall, the GWR simulations showed minor variation prior to 2041, and the main changes occurred during the last two future periods, 2041–2070 and 2071–2100. Therefore, only these two future periods will be considered in the rest of the analysis.

Table 5. Mean annual precipitation (ΔP) and temperature (ΔT) changes between 30-year periods associated with significant changes in groundwater recharge (ΔGWR) for the 12 climate scenarios and eight watersheds (W1 to W8) (cell color represents the direction of the ΔGWR: orange for decrease and blue for increase, cells are empty when ΔGWR was not significant).

Climate Scenario	Period **	W1		W2		W3		W4		W5		W6		W7		W8	
		ΔP	ΔT	ΔP	ΔT	ΔP	ΔT	ΔP	ΔT	ΔP	ΔT	ΔP	ΔT	ΔP	ΔT	ΔP	ΔT
A13 *	3	198	1.9	200	1.9	113	1.8	191	1.9								
BNU *	2	119	2.6	121	2.6	39	1.1	121	2.6	116	2.6	118	2.6	119	2.6	40	1.1
CE2	3	−45	0.7	−47	0.7	34	1.6	−47	0.7	−36	0.7	−30	0.7			13	1.6
CMS	3	180	0.9	183	0.9	−39	1.0	181	0.9	179	1.0	184	1.0	188	1.0	−38	1.0
GF3	2	92	2.1														
GF3	3	−57	0.7	−60	0.7	79	2.1	−63	0.7	−60	0.8	−48	0.8	−51	0.8	61	2.3
INM	1	36	0.2														
MIC *	2	176	2.8	182	2.8	68	2.2			158	2.8	145	2.7	176	2.8		
MIE	2	−11	1.6	−16	1.6	94	1.7	−18	1.6	−4	1.5	8	1.5	−15	1.6	60	1.7
MIE	3							−6	1.2	16	1.2	35	1.2			25	1.4
MRE *	2											117	1.5	120	1.5		
MRE *	3	135	1.9	144	1.9	88	1.5	146	1.9	127	1.9	100	1.9	88	1.9	104	1.5

* Represents RCP8.5 (scenarios without an * represent RCP4.5). ** Period 1 is 2011–2040, Period 2 is 2041–2070, and Period 3 is 2071–2100.

3.3. Inter-Annual Changes in Groundwater Recharge

The ΔGWR between the two future periods and the reference period for each climate scenario can be represented as a function of different variables (Figure 5). For each scenario, significant ΔGWR < 0 was associated with ΔP < +150 mm for all watersheds except W1 (Figure 5a,b). Inversely, significant ΔGWR > 0 was obtained when ΔP > +150 mm. The few scenarios with ΔP < 0 mm always led to ΔGWR < 0 (some significant, some not; Figure 5b). All the significant ΔGWR < 0 were simulated for +3 °C < ΔT < +5 °C, while significant ΔGWR > 0 were obtained for +2 °C < ΔT < +8 °C (Figure 5a,c). ΔGWR seemed to plateau at approximately +30 mm for both ΔT > +4.5 °C (Figure 5b,c,h, note triangle markers) and ΔT_{CM} > +6 °C (Figure 5f, December to March). Using ΔT_{CM} and ΔP_{CM} showed that ΔGWR < 0 occurred with ΔP_{CM} < +25 mm and +3 °C < ΔT_{CM} < +5 °C, except for one scenario in W3 and W8 (Figure 5d–f). All significant ΔGWR were simulated with ΔT_{CM} > +3 °C (Figure 5f). Significant ΔGWR < 0 were systematically associated with change in cold month GWR (ΔGWR_{CM}) < +25 mm and inversely for ΔGWR > 0 (Figure 5g). Significant ΔGWR < 0 were simulated with scenarios of limited changes in annual simulated AET (+50 mm < ΔAET < +95 mm), while ΔAET > +120 mm were associated with limited GWR increase (plateau around +30 mm) and a ΔT > +4.5 °C (Figure 5h). All significant annual ΔGWR were < −15 or > +15 mm (Figure 5b,c,e–h).

Of the 96 GWR simulations, 20 produced a statistically significant ΔGWR between the 1981–2010 and 2041–2070 periods, including 11 based on RCP8.5, and 39 between the 1981–2010 and 2071–2100 periods, including 21 based on RCP8.5 (Table 6). Although scenarios based on RCP8.5 did not always produce significant ΔGWR, they were more likely to produce significant ΔGWR than those based on RCP4.5. The greater number of significant changes simulated for the 2071–2100 period in comparison to the 2041–2070 period confirmed that GWR was more affected with more pronounced climate changes, be it through greater emissions or longer progression.

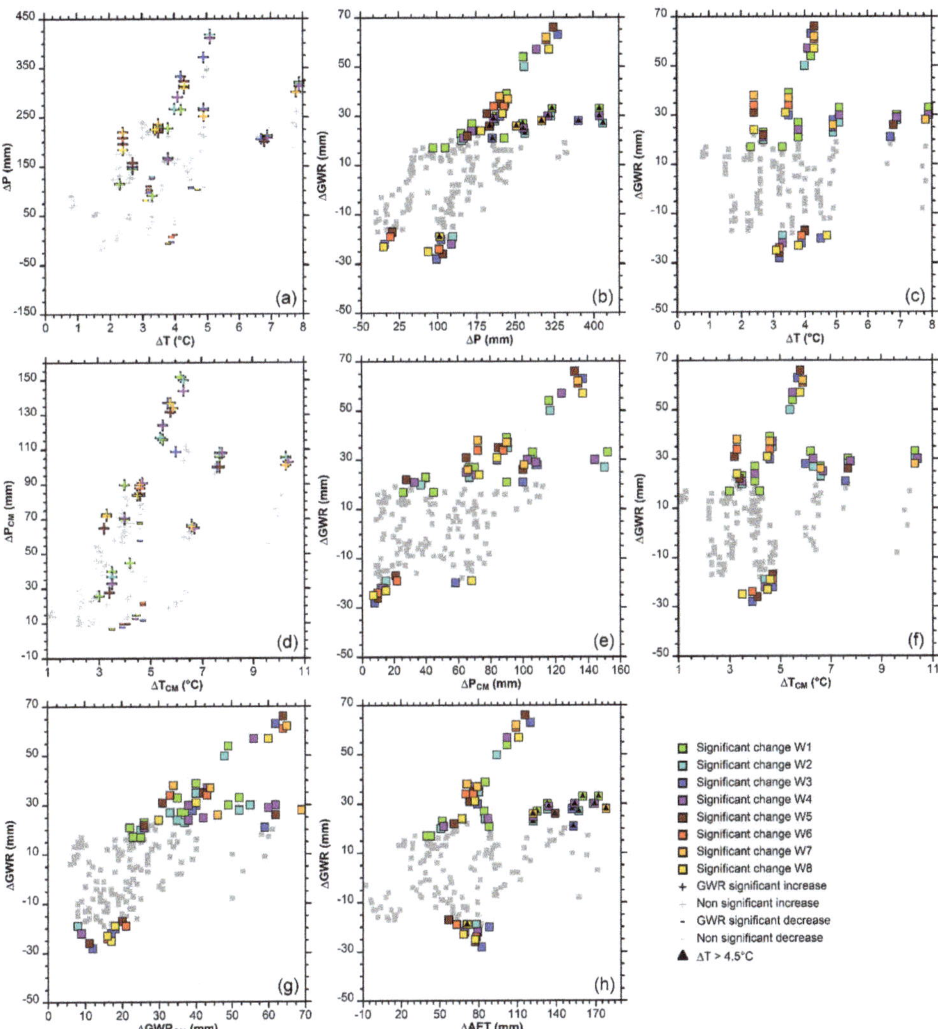

Figure 5. Changes in annual groundwater recharge (ΔGWR) between the reference period (1981–2010) and the 2041–2070 and 2071–2100 periods as a function of (**a**) changes in mean annual temperature (ΔT) and annual precipitation (ΔP), (**b**) annual precipitation changes (ΔP), (**c**) mean annual temperature changes (ΔT), (**d**) mean cold month temperature changes (ΔT$_{CM}$) and cold month precipitation changes (ΔP$_{CM}$), (**e**) cold month precipitation changes (ΔP$_{CM}$), (**f**) mean cold month temperature changes (ΔT$_{CM}$), (**g**) cold month groundwater recharge changes (ΔGWR$_{CM}$), and (**h**) annual actual evapotranspiration changes (ΔAET).

Table 6. Number of simulations with significant changes in groundwater recharge between the 1981–2010 reference period and the future periods for the eight watersheds (W1 to W8); the number of scenarios based on RCP8.5 producing significant changes is indicated in brackets.

	W1	**W2**	**W3**	**W4**	**W5**	**W6**	**W7**	**W8**
2041–2070	5 [2]	3 [2]	1 [0]	3 [2]	2 [1]	2 [1]	2 [2]	2 [1]
2071–2100	6 [4]	6 [4]	6 [3]	6 [4]	5 [2]	3 [1]	3 [2]	4 [1]

3.4. Spatial Changes in Groundwater Recharge over Time

The changes in future GWR (future periods vs. the reference period) were analyzed spatially on a cell-by-cell basis with the ensemble of scenarios (Figures 6 and 7). For the months of January, February, and March, and to a lesser extent for the month of December, significant ΔGWR > 0 was simulated for all watersheds, between +1 and >+5 mm for the 2041–2070 period, and mainly >+5 mm for the 2071–2100 period, as well as between +1 and +5 mm for December for the two periods. Although half of the changes were not significant in April for the two future periods, a clear pattern appeared during that month, with −5 mm < ΔGWR < −1 mm in the western portion of the study area and +1 mm < ΔGWR < +5 mm in the eastern portion. In May and June, significant ΔGWR < 0 was simulated, which was lower eastward and for the 2071–2100 period (locally < −5 mm). Generalized significant decreases of −5 mm < ΔGWR < −1 mm were simulated for July, August, and September for the two future periods. The ΔGWR was mainly between −5 and −1 mm from July to November for the two future periods. Non-significant ΔGWR < 0 was simulated in these months in the western and central portions of the study area. Significant ΔGWR < −5 mm was also simulated in the eastern portion for the two future periods in October and to a lesser extent in November.

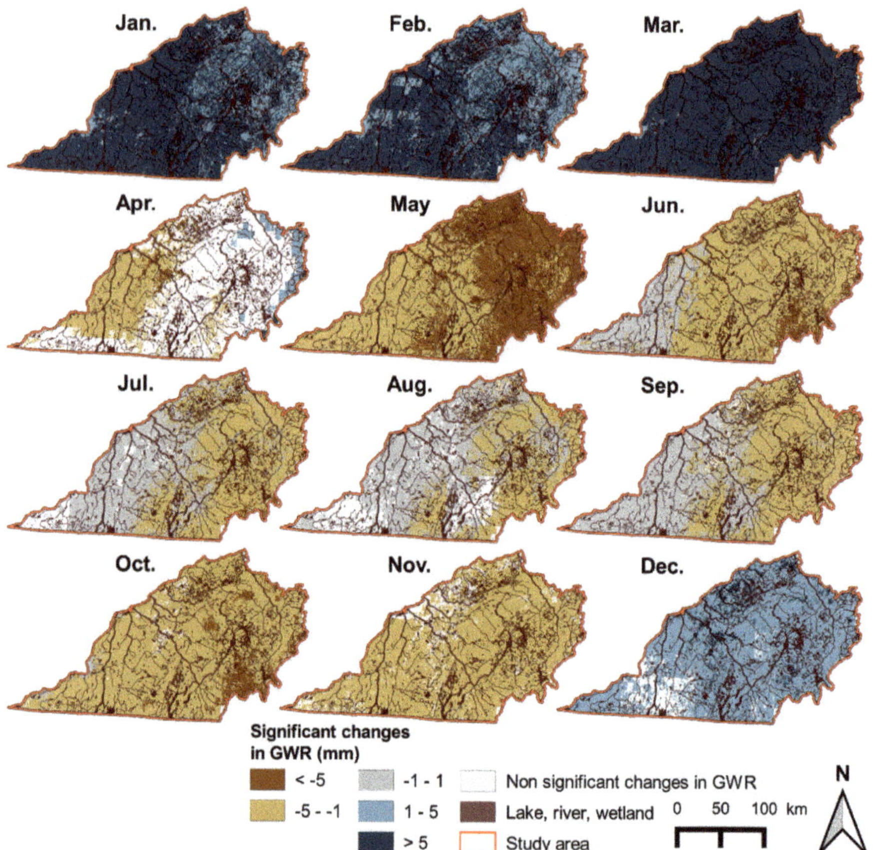

Figure 6. Spatial changes in average monthly groundwater recharge (GWR) between the reference period (1981–2010) and the 2041–2070 period for the 12 climate scenarios.

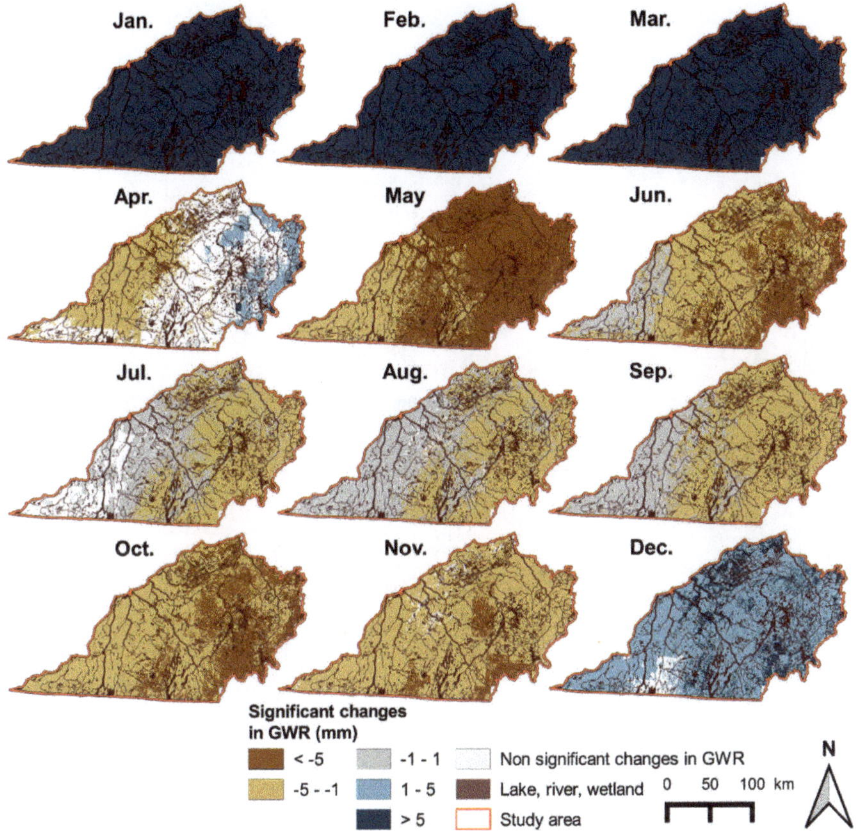

Figure 7. Spatial changes in average monthly groundwater recharge (GWR) between the reference period (1981–2010) and the 2071–2100 period for the 12 climate scenarios.

3.5. Monthly Groundwater Recharge Changes over Time

The watershed-scale monthly GWR for each period showed significant ΔGWR > 0 simulated for the eight watersheds in December, January, February, and March, with significant increases from 2041–2070 to 2071–2100 from January to March (Figure 8). The range of the ensemble changes for these months also increased remarkably in the future periods in comparison to the reference period. In April, the GWR changes were smaller and the range of the ensemble was smaller. They were mainly significant in the watersheds that are partially located in the USA (W2 and W4). The future GWR in January, February, and March exceeded that of April, which exhibited a peak during the reference period. This can already be noted in the western watersheds (W1 to W4) for the 2041–2070 period and reached similar values in the eastern watersheds (W5 to W8) in March of the 2071–2100 period. Significant ΔGWR < 0 were simulated in May and June for the two future periods and between the two future periods for all watersheds except W1 and W2. Significant ΔGWR < 0 were also simulated in July, August, September, and October for the eight watersheds and between the two future periods in October for the western watersheds (W5 to W8). From May to October, the range of changes of the ensemble was clearly smaller for all watersheds when comparing the reference period with the future periods. While the future GWR was close to zero as early as June and as late as October for the two most western watersheds (W1 and W2), the future GWR reached near-zero values between July and September in the other watersheds. Finally, significant ΔGWR < 0 was simulated for the eight watersheds in November, again with a smaller range of changes than during the reference period.

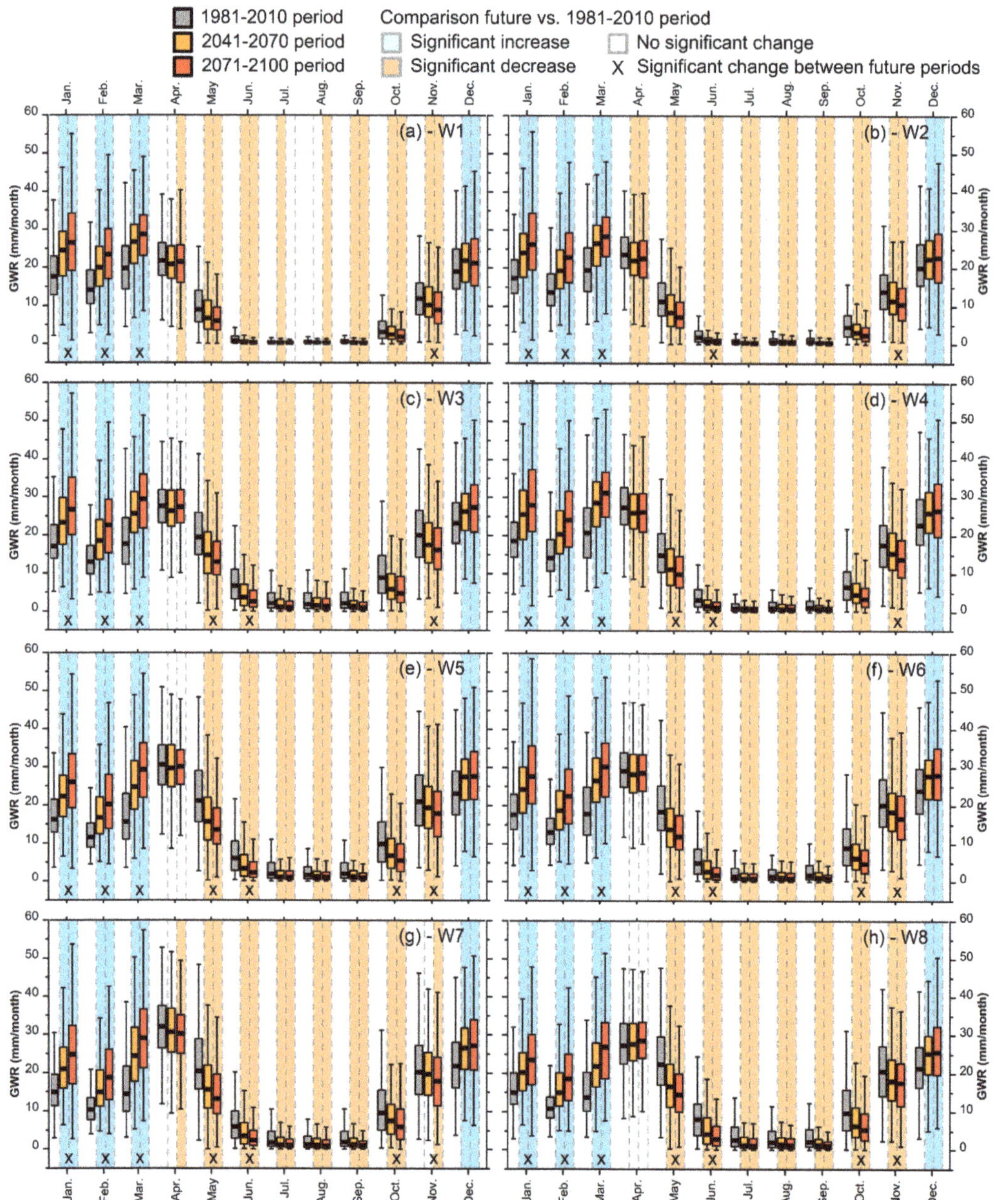

Figure 8. Monthly groundwater recharge (GWR) for the reference period (1981–2010) and the two future periods (2041–2070 and 2071–2100) for (**a**) W1 to (**h**) W8.

The sum of GWR from December to March increased by +32 mm on average (mean of the ensemble of scenarios), from +27 mm in W1 to +36 mm in W7 and between the

1981–2010 and 2071–2100 periods (Table 4). The sum of the GWR from May to November (months with T > 0 °C) decreased by −19 mm, from −9 mm in W1 to −26 mm in W8 and between the 1981–2010 and 2071–2100 periods. These seasonal changes were within the range of uncertainty of the annual GWR.

4. Discussion

4.1. Future Groundwater Recharge Dynamics

For all watersheds except W1, the GWR for the 2011–2040 period was not statistically different from that of the 1981–2010 period. Of the eight watersheds, significant GWR changes occurred with two to four climate scenarios between the 2011–2040 and 2041–2070 horizons and with four to seven climate scenarios between the 2041–2070 and 2071–2100 horizons (Figure 4). For this reason, the results were compared only between the 2041–2070 and 2071–2100 periods and the 1981–2010 period.

The simulations showed both increases and decreases in GWR in the future, hence markedly increasing the range of possible future conditions from those simulated in the reference period (Figure 4). The climate scenarios based on RCP8.5 were the wettest (140 mm < ΔP < 220 mm, Figure 2) and thus produced increasing GWR rates in the future. Other studies have observed a wide range of hydrological responses to climate change in cold and humid regions or regions with snow-dependent hydrology [2,11,14]. Kurylyk and MacQuarrie [38] simulated increased future annual GWR under four climate scenarios and decreased GWR under three climate scenarios in New Brunswick (eastern Canada). Guay et al. [41] have shown, based on the simulation of future river flows in 305 watersheds in Quebec under 87 climate scenarios, that it was unclear whether future annual river flows would increase or decrease by the 2041–2070 period. Inversely, Sulis et al. [52] mainly simulated decreased future GWR in part of the Chateaugay River watershed (W1) with the integrated CATHY model for the 2041–2065 period (increase under one scenario). These authors used 12 climate scenarios based on the high-emission SRES A2 greenhouse gas projections, with annual precipitation increases of close to 0 to +20% between the future and the reference periods. Differences in future GWR depend on the choice of future horizon and emission scenarios, as well as on the type of model used to derive the GWR [15].

The analysis of monthly recharge allowed major shifts in the intra-annual changes in future GWR to be identified. The results show that winter GWR could significantly increase due to warmer winters and lead to an earlier spring GWR peak. Other studies in eastern Canada have obtained similar results [30,33,37–39]. Similarly, Grinevskiy et al. [53] simulated GWR with an unsaturated zone model (HYDRUS-1D) in 22 sites spread over western Russia (humid climate, cold in the north, temperate in the south) and observed increased GWR during winter, which was linked to wetter and warmer winters in the North, but not in the South of the study area. Such results are reported for cold and humid climates and regions of snow-dominated hydrology with more available liquid water during winter, which is linked to warmer temperatures that affect not only GWR, but the entire hydrologic dynamic [2,11,27,29,34,35,41,54]. These future GWR conditions are supported by observations of past groundwater level time series showing a similar shift in the GWR peak from spring (snowmelt) to winter (rain) in Fennoscandia (Northern Europe, transition between temperate and cold climates) associated with a warming climate between the 1980–1989 and 2001–2010 periods [4].

In the present study, the GWR scenarios showed a statistically significant decrease from May to November (Figure 8). The future GWR was close to zero from July to September (similarly to the reference period), except in the western and warmest watersheds, W1 and W2, where the low flow period began a month earlier (June) and ended a month later (October). Similar results were obtained in different cold and humid climates or regions of snow-dominated hydrology. Guay et al. [41] noticed small or negligible changes in river flows during the summer, a period of the year where flow rates are already very low, extending until October. Expected dryer summer low flow rates were also reported by Addor et al. [11] and Arnoux et al. [27] for the Swiss Alps and by Dieraurer et al. [34] for watersheds across the Rockies (western North America), and they were linked to reduced snowpack, leading to limited snowmelt contribution to spring flows. In this study, the GWR scenarios were similar for all watersheds during summer, thus increasing the certainty of the expected summer decrease. Despite the high uncertainty in simulations of future hydrologic conditions, Addor et al. [11] reported a good convergence of results toward lower summer flow (90% of the scenarios) for Swiss alpine watersheds, similar to that of the current study. However, Arnoux et al. [27] showed that post-glacial Quaternary deposits can contribute to the mitigation of the impact of climate change on summer low flows in alpine catchments due to their water storage capacity, which supports river low flows during long dry spells. This is an indication that water-bearing unconsolidated superficial materials could be an indicator of watershed response to climate change.

Aygün et al. [2] showed that the hydrology of cold and humid regions (northern regions of North America and Eurasia outside of the permafrost zone) with near-freezing annual temperatures were more sensitive to climate change than regions with substantially colder climates. The current study showed differences in the watershed response from west to east that followed the regional temperature gradient (decrease of mean annual temperature). These findings were most likely possible because of the use of a single model across the region and a robust knowledge of the past dynamics. Larocque et al. [15] did not find such a clear trend from west to east in their review of modeling studies of climate change impacts on groundwater systems in eastern Canada.

4.2. Climate Changes Impacting Groundwater Recharge

The groundwater recharge changes became statistically significant when ΔGWR was < -15 or > +15 mm for the two future periods (Figure 5). More specifically, small GWR changes could not be interpreted as being different from the simulated variability of the 1981–2010 reference period for one to five of the 12 scenarios for the 2041–2070 period and three to six of the scenarios for the 2071–2100 period (Table 6). The increasing number of scenarios with significant changes for the 2071–2100 period is coherent with results from Goderniaux et al. [42] in the Geer Basin (Belgium), where projections of groundwater levels obtained using an ensemble of 30 climate scenarios became greater than the variability of the 1961–1990 period only in 2085. Similarly, using an ensemble of 54 climate scenarios, Addor et al. [11] demonstrated that flow rate changes in alpine catchments became significantly different from those of the 1980–2009 reference period only after the 2050 horizon. They showed that significant changes were simulated even under the climate scenarios with the lowest emissions based on RCP2.6 (not used in this study). In contrast, this study showed that climate scenarios based on RCP8.5 did not systematically produce significant changes between the future periods and the reference period, although they tended to simulate significant changes and higher future GWR than scenarios based on RCP4.5 more often. Henceforth, using a large ensemble of climate scenarios appears to be necessary to provide a representative sample of possible future precipitation and temperature.

One of the main novelties of this work lies in the identification of climate conditions leading to statistically significant changes in future GWR. Significant ΔGWR < 0 was simulated only with ΔP < +150 and ΔP$_{CM}$ < +25 mm. ΔP < 0 always led to ΔGWR < 0, but the latter was not necessarily significant. Inversely, ΔGWR > 0 were significant only with ΔP > +150 and ΔP$_{CM}$ > +25 mm. Therefore, ΔP $\approx$ +150 and ΔP$_{CM}$ $\approx$ +25 mm appear to be regional thresholds for determining the direction of future GWR changes.

Another contribution of this work was to determine that significant ΔGWR < 0 was systematically associated with ΔT and ΔT$_{CM}$ ranging between +3 and +5 °C, while significant ΔGWR > 0 was found for +2 °C < ΔT < +8 °C and +3 °C < ΔT$_{CM}$ < +11 °C. Interestingly, ΔT > +4.5 °C (or ΔT$_{CM}$ > +6 °C) led to ΔAET > +120 mm, thus limiting ΔGWR to +30 mm. Therefore, ΔT $\approx$ +2 °C and ΔT$_{CM}$ $\approx$ +3 °C appear to be regional thresholds for significant GWR changes (increase or decrease), while ΔT > +4.5 °C triggers GWR increase. These temperature thresholds control future GWR through the modification of the cold month hydrology and the evolution of ET, which also depends on the adaptation of vegetation to climate change.

This study demonstrated that, on an annual basis, $\Delta GWR_{CM} > +25$ mm compensated for decreased GWR during the rest of the year and produced statistically significant $\Delta GWR > 0$. This is coherent with numerous previous studies showing that the seasonality of the entire hydrologic dynamic (GWR, groundwater storage, groundwater level, stream flow) in cold and humid climates or regions of snow-dependent hydrology was affected by the increase in available liquid water during warmer winters, counterbalancing the decreasing availability of water during summer [2,4,11,27,30,35,39,41]. Rivard et al. [39] observed that changes in future GWR were most sensitive to winter temperature in simulations with a spatialized water budget model in Nova Scotia (HELP, eastern Canada, not overlapping the current study area). This was more due to increased amounts of liquid winter precipitation that was readily available for infiltration than to changes in precipitation amounts. Interestingly, Wright and Novakowski [33] showed that winter recharge events on a fractured bedrock (Ontario, Canada, frozen during winter) could bypass the frozen soil and reach unfrozen fractures at the soil/bedrock interface. They concluded that winter rainfall events could produce more GWR than during the rest of the year, thus making the precipitation form and amount during this period a sensitive GWR variable. The differences in these studies may be due to their respective scales. The local scale associated with GWR estimates based on well observations used in some studies [4,33] is in dire contrast to the 250 m × 250 m resolution used by Rivard et al. [39] or the 500 m × 500 m in the HydroBudget model. In addition, the sub-hourly sampling time-step of other studies [33,55] is not comparable to the daily time-steps aggregated into monthly inter-annual results presented here. Nevertheless, all the available studies for Eastern Canada confirm the importance of future winter GWR in the overall annual GWR dynamic, as well as the importance of capturing local-scale (meter order) processes in regional-scale GWR simulations.

From a different perspective, Sulis et al. [52] showed that changes in future GWR in a sub-watershed of W1 were linked to intra-annual patterns of the climate scenarios (more snowmelt during winter, less rain during the fall, the duration of successive days with daily precipitation > 1 mm/d) rather than being related to annual precipitation changes. The integrated CATHY model (daily time-step) seemed sensitive to the dryness conditions of the soil [56], thus inducing more percolation through the unsaturated zone (GWR) for climate scenarios with regular summer rainfall events than for scenarios with more intense but less frequent rainfall events. Similar conclusions were reached by Wright and Novakowski [33] at the well scale in a fractured bedrock aquifer for winter GWR events in Ontario. Finally, Rathay et al. [55] observed that increasing rainfall intensity, from <1 mm/h to > 1 mm/h, produced a decrease in the rainfall–groundwater level cross-correlation coefficients in a bedrock aquifer in the temperate climate of British Columbia (Canada). Although they did not identify a rainfall intensity threshold limiting GWR, these authors concluded that more intense rainfall events produced more surface and subsurface runoff rather than increasing GWR rates. Although these studies highlighted that precipitation intensity can be an important factor for future GWR changes in humid climates, the sensitivity of GWR to this parameter was not a focus of the current study.

4.3. Future Groundwater Recharge Simulation in Cold and Humid Climates

The clustering method used to select the subset of climate scenarios was based on ten criteria including changes in seasonal and annual precipitation, as well as changes in temperature, but did not include changes in precipitation intensity. Although recent work has projected the intensification of year-round precipitation in North America [57], precipitation intensity changes for the province of Quebec are not yet clear [25]. Further research needs to assess the impact of this variable on increasing or decreasing future GWR in cold and humid climates on intra-annual and inter-annual time scales.

Considering the range of changes in future recharge, understanding GWR under future conditions probably lies mostly in the capacity to adequately simulate GWR during the cold months, the period corresponding to the greatest changes in terms of absolute value in the study area. Stream flow or GWR simulations in cold and humid climates are sensitive to snow-related calibration parameters, such as the melting temperature and melting coefficient [22,58]. However, Melsen and Guse [59] showed that these parameters were less sensitive when simulating river flow in 605 USA watersheds under future conditions with decreased snowpack. Therefore, an evolution of the snow-related parameters could be expected under future conditions. Improving the simulation of winter GWR in cold regions will necessitate a better understanding of the roles of snow dynamics and soil frost in changing conditions, and future work should be aimed at calibrating these parameters for long-term regional-scale simulations.

The current study was based on the HB model, which was calibrated and validated over an exceptionally long period of time (57 years from 1961 to 2017), ensuring satisfying representativeness of the long-term and regional-scale hydrological dynamics [22]. The resulting GWR scenarios used constant model parameters over time under the hypothesis that the system was stationary in time and no significant land-use change occurred. However, Jaramillo et al. [60] linked a 40-year increase in AET rates of 65 mm/yr in the Stockholm region (Sweden, temperate to cold climate transition zone) to land-use change, with the massive conversion of semi-natural grasslands (mowing) to cereal and fodder harvesting at the beginning of the 20th century. For Sweden as well, Destouni et al. [61] compared the evolution of evapotranspiration (ET) and runoff (R) for nine watersheds in temperate and cold climates that remained stationary in time or were affected by hydropower and non-irrigated agricultural development during the 20th century. Despite precipitation and temperature increases, they found that ET and R remained stable in unregulated watersheds, while hydropower development increased ET and decreased R, and agriculture development increased both ET and R. These hydrological changes impact the regional water budget, and therefore most likely propagate to GWR. Alternatively, Guerrero-Morales et al. [62] found that land cover changes accounted for 25% of the GWR decrease in an urbanized watershed in western Mexico (warm and humid climate) under climate change conditions by the 2050 horizon. Although Kløve et al. [9] and Taylor et al. [10] stated that climate change studies should consider land-use change, integrating land-use scenarios into future GWR simulations in cold and humid climates has not been widely reported in the scientific literature. Further study of this important question could lead to the identification of other factors than climate that determine the extent of possible GWR changes. Considering land-use change would also probably increase the uncertainty of future GWR simulations [63].

Reinecke et al. [17] concluded on the importance of coupling biosphere dynamic simulations to long-term GWR simulation, especially at the global scale, where increases in atmospheric CO_2 concentrations could lead to more active vegetation, which would, in turn, impact GWR estimates. Koirala et al. [64] showed that vegetation had a large impact on the water budget through AET, especially in humid climates. To avoid using scenarios of future solar radiation and other climate variables that are less readily available, more difficult to bias-correct, and may introduce additional uncertainty compared to the more common temperature and precipitation scenarios [25], the maximum daily ET in HB was based on the simple formula from Oudin et al. [47], which only used daily temperature, latitude, and Julian day (as a proxy for extraterrestrial radiation). However, considering the regional scale of the study and the long-term simulation period, more work should be dedicated to improving AET simulations for cold and humid climates, especially considering the uncertainty related to plant adaptation to warmer climates. This could impact the temperature thresholds identified in this study. Specific AET calibration could be developed using spatialized time-series of the measured AET, or the impact of coupling biosphere dynamics and GWR at the regional scale could be tested.

4.4. Using These Results for Adaptation

Studies clearly show that GWR in cold and humid climates could follow different paths of change depending on specific climate conditions, geology, morphology, and land use [15,38]. Taking into consideration uncertainty in future climate conditions is another major challenge, as this study showed with future GWR that can either increase or decrease at the regional scale depending on the climate scenario It is thus extremely difficult to provide concrete recommendations to water managers despite the increasing body of knowledge [14].

Nevertheless, several patterns in the future evolution of GWR emerged with a relatively high level of confidence. For example, the significant projected decrease in GWR from May to November as soon as the 2041–2070 period and the substantial increase in GWR from December to March clearly stand out. A cold month GWR increase of > +25 mm will compensate for the decrease throughout the rest of the year, suggesting stable groundwater resources. Additionally, this work provides threshold values for changes in precipitation and temperature that lead to likely increases or decreases in future GWR (Figure 9). These thresholds could be used in integrated water resource management plans, where they could trigger specific actions (e.g., if local warming reaches 1.5, 2, or 3 °C, associated with stable precipitation increase of +50 or +100 mm). Although they would probably be similar in other cold and humid climates in post-glacial geological environments, these thresholds will need to be tested in different contexts.

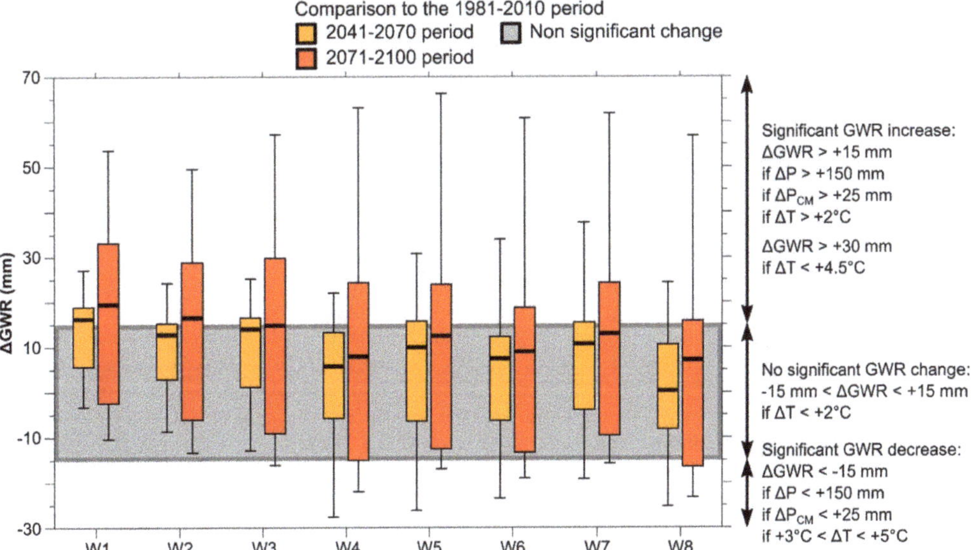

Figure 9. Annual groundwater recharge changes (ΔGWR) between the reference period (1981–2010) and future periods (2041–2070 and 2071–2100) for the eight watersheds (W1–W8) and 12 scenarios. The associated precipitation and temperature thresholds are displayed on the right, and the gray zone indicates the −15 to +15 mm non-significant change range in GWR. CM stands for cold months (December to March).

In cold and humid climates, GWR generally represents the actual aquifer renewal rates—the total flow discharging to superficial water bodies [4,19,21]. A decrease in future GWR from May to November means that groundwater inflow into superficial water bodies and groundwater levels will decrease when water demand for drinking water, agriculture, industrial purposes, hydroelectricity, and recreation is the highest [19,65] and when river flows come almost exclusively from a connected aquifer. Considering the high confidence in the simulation of the decreasing GWR from May to November, it is expected that water use conflicts will increase in future decades.

The identified thresholds were related to potential GWR, i.e., the maximum GWR that can reach the saturated zone [22]. Future changes in actual GWR are expected to be closely linked to those in potential GWR. For scenarios and periods where potential GWR is expected to decrease, actual GWR will most likely decrease as well due to the reduction of available water. Inversely, for expected increases in potential GWR, actual GWR changes would vary depending on the AET rates. Future work studying the propagation of these changes should focus on the periods of expected potential GWR increases.

5. Conclusions

In cold and humid climates, the impact of climate change will propagate in groundwater systems and more broadly to regional hydrologic dynamics through GWR. Estimates of changes in GWR under future climate conditions are therefore strategic for long-term water resource management. This work has provided new data for assessing climate change impacts on GWR and to identify controlling processes and thresholds for cold and humid climates. One of the outcomes was the simulation of the first set of 12 transient regional-scale GWR scenarios for the 1951–2100 period in southern Quebec. Simulated using a water budget model and a set of 12 climate scenarios maximizing the future climate variability (12 GCMs using RCP4.5 and RCP8.5), the spatio-temporal GWR scenarios showed notable changes occurring in the 2041–2070 and 2071–2100 periods. Warming temperatures were between +1 and +5 °C at the 2041–2070 horizon (in comparison with the 1981–2010 reference period), and the precipitation change pattern was more variable, including an increase of +10% to +80% in the available liquid water between March and December. Increasing and decreasing annual GWR was simulated. However, major impacts were found in the monthly dynamics, with a statistically significant decrease in future GWR from May to November compensated by a statistically significant increase in future GWR from December to March. The periods of null or very low GWR rates were lengthened by one month in June and October for the warmer watersheds. Overall, the average annual GWR change was positive if the increase in future cold month GWR was higher than +25 mm, offsetting the decrease for the rest of the year. Such results were coherent with previous findings in other regions of cold and humid climates.

The novelty of this work lies in linking changing climate conditions to the direction and amplitude of statistically significant changes in future regional GWR through specific precipitation and temperature change thresholds. All significant changes in GWR were >+15 or <−15 mm/yr and were only produced by warming temperatures >+2 °C. A significant decrease in future GWR was always simulated under future increases in annual precipitation of <+150 mm and cold month precipitation changes of <+25 mm, along with warming temperatures of between +3 and +5 °C (for annual and cold months). A significant increase in future GWR was systematically simulated under increases in annual precipitation of >+150 mm and cold month precipitation increases of >+25 mm, along with warming temperatures of >+2 °C. A future temperature increase of >+4.5 °C produced more intense AET rates, thus limiting the increase in future GWR to approximately +30 mm, irrespective of the precipitation increase. These thresholds are sufficiently straightforward for general use and for integrated water resource management plans.

Author Contributions: All authors contributed to writing the manuscript. E.D., M.L., and S.G. developed the approach. M.B. selected climate scenarios through cluster analysis. Simulations and figure preparation were done by E.D., M.L. obtained the research grant and supervised the research. All authors have read and agreed to the published version of the manuscript.

Funding: This research was funded by the Quebec Ministry of Environment and Climate Change (Ministère de l'Environnement et de la Lutte contre les changements climatiques).

Institutional Review Board Statement: Not applicable.

Informed Consent Statement: Not applicable.

Data Availability Statement: The simulated scenarios presented in this work are available here: https://doi.org/10.5683/SP3/SWH4O1.

Acknowledgments: The authors are grateful to the Ouranos Consortium for providing downscaled climate scenarios and acknowledge the model output data from the World Climate Research Programme's Coupled Modelling Intercomparison Project Phase 5 (CMIP5) as well as the gridded observation data made available by Natural Resources Canada's (NRCan).

Conflicts of Interest: The authors declare no conflict of interest. The funders had no role in the design of the study; in the collection, analyses, or interpretation of data; in the writing of the manuscript, or in the decision to publish the results.

References

1. Gudmundsson, L.; Boulange, J.; Do, H.X.; Gosling, S.N.; Grillakis, M.G.; Koutroulis, A.G.; Leonard, M.; Liu, J.; Schmied, H.M.; Papadimitriou, L.; et al. Globally observed trends in mean and extreme river flow attributed to climate change. *Science* **2021**, *371*, 1159–1162. [CrossRef]
2. Aygün, O.; Kinnard, C.; Campeau, S. Impacts of climate change on the hydrology of northern midlatitude cold regions. *Prog. Phys. Geogr. Earth Environ.* **2019**, *44*, 338–375. [CrossRef]
3. Lee, B.; Hamm, S.-Y.; Jang, S.; Cheong, J.-Y.; Kim, G.-B. Relationship between groundwater and climate change in South Korea. *Geosci. J.* **2013**, *18*, 209–218. [CrossRef]
4. Nygren, M.; Giese, M.; Kløve, B.; Haaf, E.; Rossi, P.M.; Barthel, R. Changes in seasonality of groundwater level fluctuations in a temperate-cold climate transition zone. *J. Hydrol. X* **2020**, *8*, 100062. [CrossRef]
5. Arctic Climate Impact Assessment. *Impacts of a Warming Arctic*; Cambridge University Press: Cambridge, UK; New York, NY, USA, 2004; ISBN 978-0-521-61778-9.
6. Döll, P. Vulnerability to the impact of climate change on renewable groundwater resources: A global-scale assessment. *Environ. Res. Lett.* **2009**, *4*, 035006. [CrossRef]
7. Döll, P.; Trautmann, T.; Gerten, D.; Schmied, H.M.; Ostberg, S.; Saaed, F.; Schleussner, C.-F. Risks for the global freshwater system at 1.5 °C and 2 °C global warming. *Environ. Res. Lett.* **2018**, *13*, 044038. [CrossRef]
8. Green, T.R.; Taniguchi, M.; Kooi, H.; Gurdak, J.J.; Allen, D.M.; Hiscock, K.M.; Treidel, H.; Aureli, A. Beneath the surface of global change: Impacts of climate change on groundwater. *J. Hydrol.* **2011**, *405*, 532–560. [CrossRef]
9. Klöve, B.; Ala-Aho, P.; Bertrand, G.; Gurdak, J.J.; Kupfersberger, H.; Kværner, J.; Muotka, T.; Mykrä, H.; Preda, E.; Rossi, P.M.; et al. Climate change impacts on groundwater and dependent ecosystems. *J. Hydrol.* **2014**, *518*, 250–266. [CrossRef]
10. Taylor, R.G.; Scanlon, B.; Döll, P.; Rodell, M.; Van Beek, R.; Wada, Y.; Longuevergne, L.; Leblanc, M.; Famiglietti, J.S.; Edmunds, M.; et al. Ground water and climate change. *Nat. Clim. Chang.* **2013**, *3*, 322–329. [CrossRef]
11. Addor, N.; Rössler, O.; Köplin, N.; Huss, M.; Weingartner, R.; Seibert, J. Robust changes and sources of uncertainty in the projected hydrological regimes of Swiss catchments. *Water Resour. Res.* **2014**, *50*, 7541–7562. [CrossRef]
12. Allen, D.M.; Cannon, A.; Toews, M.; Scibek, J. Variability in simulated recharge using different GCMs. *Water Resour. Res.* **2010**, *46*, W00F03. [CrossRef]
13. Crosbie, R.S.; Pickett, T.; Mpelasoka, F.S.; Hodgson, G.; Charles, S.P.; Barron, O.V. An assessment of the climate change impacts on groundwater recharge at a continental scale using a probabilistic approach with an ensemble of GCMs. *Clim. Chang.* **2012**, *117*, 41–53. [CrossRef]
14. Smerdon, B.D. A synopsis of climate change effects on groundwater recharge. *J. Hydrol.* **2017**, *555*, 125–128. [CrossRef]
15. Larocque, M.; Levison, J.; Martin, A.; Chaumont, D. A review of simulated climate change impacts on groundwater resources in Eastern Canada. *Can. Water Resour. J. Rev. Can. Des Ressour. Hydr.* **2019**, *44*, 22–41. [CrossRef]
16. Moeck, C.; Brunner, P.; Hunkeler, D. The influence of model structure on groundwater recharge rates in climate-change impact studies. *Hydrogeol. J.* **2016**, *24*, 1171–1184. [CrossRef]
17. Reinecke, R.; Schmied, H.M.; Trautmann, T.; Andersen, L.S.; Burek, P.; Flörke, M.; Gosling, S.N.; Grillakis, M.; Hanasaki, N.; Koutroulis, A.; et al. Uncertainty of simulated groundwater recharge at different global warming levels: A global-scale multi-model ensemble study. *Hydrol. Earth Syst. Sci.* **2021**, *25*, 787–810. [CrossRef]
18. Gleeson, T.; Marklund, L.; Smith, L.; Manning, A. Classifying the water table at regional to continental scales. *Geophys. Res. Lett.* **2011**, *38*, L05401. [CrossRef]
19. Kløve, B.; Kvitsand, H.M.L.; Pitkänen, T.; Gunnarsdottir, M.J.; Gaut, S.; Gardarsson, S.; Rossi, P.M.; Miettinen, I. Overview of groundwater sources and water-supply systems, and associated microbial pollution, in Finland, Norway and Iceland. *Appl. Hydrogeol.* **2017**, *25*, 1033–1044. [CrossRef]
20. Meyzonnat, G.; Barbecot, F.; Alazard, M.; McCormack, R. La Richesse de La Ressource En Eau Du Québec (Richness of Quebec's Water Ressource). *Géologues Rev. Off. Société Géologique Fr.* **2018**, *198*, 69–75.
21. Rivera, A. *Canada's Groundwater Resources*; Fitzhenry & Whiteside: Markham, ON, Canada, 2014; ISBN 978-1-55455-292-4.

22. Dubois, E.; Larocque, M.; Gagné, S.; Meyzonnat, G. Simulation of long-term spatiotemporal variations in regional-scale groundwater recharge: Contributions of a water budget approach in cold and humid climates. *Hydrol. Earth Syst. Sci.* **2021**, *25*, 6567–6589. [CrossRef]

23. Cuthbert, M.O.; Gleeson, T.; Moosdorf, N.; Befus, K.M.; Schneider, A.; Hartmann, J.; Lehner, B. Global patterns and dynamics of climate–groundwater interactions. *Nat. Clim. Chang.* **2019**, *9*, 137–141. [CrossRef]

24. Collins, M.; Knutti, R.; Arblaster, J.; Dufresne, J.-L.; Fichefet, T.; Friedlingstein, P.; Gao, X.; Gutowski, W.J.; Johns, T.; Krinner, G.; et al. Long-term Climate Change: Projections, Commitments and Irreversibility. In *Climate Change 2013: The Physical Science Basis. Contribution of Working Group I to the Fifth Assessment Report of the Intergovernmental Panel on Climate Change*; Stocker, T.F., Qin, D., Plattner, G.-K., Tignor, M., Allen, S.K., Boschung, J., Nauels, A., Xia, Y., Bex, V., Midgley, P.M., Eds.; Cambridge University Press: Cambridge, UK; New York, NY, USA, 2013; pp. 1029–1136. ISBN 978-1-107-66182-0.

25. Ouranos. *Vers l'adaptation—Synthèse Des Connaissances Sur Les Changements Climatiques Au Québec. (Toward Adaptation—Synthesis on the Knowledge about Climate Change in Quebec)*, 2015 ed.; Ouranos: Montréal, QC, Canada, 2015; ISBN 978-2-923292-18-2.

26. Vincent, L.; Zhang, X.; Mekis, E.; Wan, H.; Bush, E. Changes in Canada's Climate: Trends in Indices Based on Daily Temperature and Precipitation Data. *Atmosphere-Ocean* **2018**, *56*, 332–349. [CrossRef]

27. Arnoux, M.; Brunner, P.; Schaefli, B.; Mott, R.; Cochand, F.; Hunkeler, D. Low-flow behavior of alpine catchments with varying quaternary cover under current and future climatic conditions. *J. Hydrol.* **2021**, *592*, 125591. [CrossRef]

28. Barnett, T.P.; Adam, J.C.; Lettenmaier, D.P. Potential impacts of a warming climate on water availability in snow-dominated regions. *Nature* **2005**, *438*, 303–309. [CrossRef] [PubMed]

29. Berghuijs, W.R.; Woods, R.A.; Hrachowitz, M. A precipitation shift from snow towards rain leads to a decrease in streamflow. *Nat. Clim. Chang.* **2014**, *4*, 583–586. [CrossRef]

30. Cochand, F.; Therrien, R.; Lemieux, J.-M. Integrated Hydrological Modeling of Climate Change Impacts in a Snow-Influenced Catchment. *Groundwater* **2018**, *57*, 3–20. [CrossRef]

31. Hayashi, M.; Farrow, C.R. Watershed-scale response of groundwater recharge to inter-annual and inter-decadal variability in precipitation (Alberta, Canada). *Hydrogeol. J.* **2014**, *22*, 1825–1839. [CrossRef]

32. Scibek, J.; Allen, D.M.; Cannon, A.; Whitfield, P. Groundwater–surface water interaction under scenarios of climate change using a high-resolution transient groundwater model. *J. Hydrol.* **2007**, *333*, 165–181. [CrossRef]

33. Wright, S.N.; Novakowski, K.S. Impacts of warming winters on recharge in a seasonally frozen bedrock aquifer. *J. Hydrol.* **2020**, *590*, 125352. [CrossRef]

34. Dierauer, J.R.; Whitfield, P.H.; Allen, D.M. Climate Controls on Runoff and Low Flows in Mountain Catchments of Western North America. *Water Resour. Res.* **2018**, *54*, 7495–7510. [CrossRef]

35. Jenicek, M.; Seibert, J.; Zappa, M.; Staudinger, M.; Jonas, T. Importance of maximum snow accumulation for summer low flows in humid catchments. *Hydrol. Earth Syst. Sci.* **2016**, *20*, 859–874. [CrossRef]

36. Kong, Y.; Wang, C.-H. Responses and changes in the permafrost and snow water equivalent in the Northern Hemisphere under a scenario of 1.5 °C warming. *Adv. Clim. Chang. Res.* **2017**, *8*, 235–244. [CrossRef]

37. Jyrkama, M.I.; Sykes, J.F. The impact of climate change on spatially varying groundwater recharge in the grand river watershed (Ontario). *J. Hydrol.* **2007**, *338*, 237–250. [CrossRef]

38. Kurylyk, B.L.; MacQuarrie, K.T. The uncertainty associated with estimating future groundwater recharge: A summary of recent research and an example from a small unconfined aquifer in a northern humid-continental climate. *J. Hydrol.* **2013**, *492*, 244–253. [CrossRef]

39. Rivard, C.; Paniconi, C.; Vigneault, H.; Chaumont, D. A watershed-scale study of climate change impacts on groundwater recharge (Annapolis Valley, Nova Scotia, Canada). *Hydrol. Sci. J.* **2014**, *59*, 1437–1456. [CrossRef]

40. Direction De L'expertise Hydrique (DEH) Atlas Hydroclimatique Du Québec Méridional (Hydroclimatic Atlas of Southern Quebec). Available online: https://www.cehq.gouv.qc.ca/atlas-hydroclimatique/Hydraulicite/Qmoy.htm (accessed on 13 April 2021).

41. Guay, C.; Minville, M.; Braun, M. A global portrait of hydrological changes at the 2050 horizon for the province of Québec. *Can. Water Resour. J. Rev. Can. Des Ressour. Hydr.* **2015**, *40*, 285–302. [CrossRef]

42. Goderniaux, P.; Brouyère, S.; Blenkinsop, S.; Burton, A.; Fowler, H.; Orban, P.; Dassargues, A. Modeling climate change impacts on groundwater resources using transient stochastic climatic scenarios. *Water Resour. Res.* **2011**, *47*, W12516. [CrossRef]

43. Hund, S.V.; Allen, D.M.; Morillas, L.; Johnson, M.S. Groundwater recharge indicator as tool for decision makers to increase socio-hydrological resilience to seasonal drought. *J. Hydrol.* **2018**, *563*, 1119–1134. [CrossRef]

44. Bertrand, G.; Ponçot, A.; Pohl, B.; Lhosmot, A.; Steinmann, M.; Johannet, A.; Pinel, S.; Caldirak, H.; Artigue, G.; Binet, P.; et al. Statistical hydrology for evaluating peatland water table sensitivity to simple environmental variables and climate changes application to the mid-latitude/altitude Frasne peatland (Jura Mountains, France). *Sci. Total. Environ.* **2020**, *754*, 141931. [CrossRef]

45. Dubois, E.; Larocque, M.; Gagné, S.; Meyzonnat, G. *HydroBudget User Guide: Version 1.2*; Université du Québec à Montréal, Département des sciences de la Terre et de l'atmosphère: Montréal, QC, Canada, 2021; Available online: https://archipel.uqam.ca/14075/ (accessed on 14 November 2021).

46. Dubois, E.; Larocque, M.; Gagne, S.; Meyzonnat, G. HydroBudget—Groundwater Recharge Model in R. Dataverse [Code]. 2021. Available online: https://doi.org/10.5683/SP3/EUDV3H (accessed on 14 November 2021).

47. Oudin, L.; Hervieu, F.; Michel, C.; Perrin, C.; Andréassian, V.; Anctil, F.; Loumagne, C. Which potential evapotranspiration input for a lumped rainfall–runoff model? Part 2—Towards a simple and efficient potential evapotranspiration model for rainfall–runoff modelling. *J. Hydrol.* **2005**, *303*, 290–306. [CrossRef]

48. Casajus, N.; Perie, C.; Logan, T.; Lambert, M.-C.; De Blois, S.; Berteaux, D. An Objective Approach to Select Climate Scenarios when Projecting Species Distribution under Climate Change. *PLoS ONE* **2016**, *11*, e0152495. [CrossRef]

49. Hopkinson, R.F.; McKenney, D.W.; Milewska, E.J.; Hutchinson, M.F.; Papadopol, P.; Vincent, L.A. Impact of Aligning Climatological Day on Gridding Daily Maximum–Minimum Temperature and Precipitation over Canada. *J. Appl. Meteorol. Clim.* **2011**, *50*, 1654–1665. [CrossRef]

50. Hutchinson, M.F.; McKenney, D.W.; Lawrence, K.; Pedlar, J.H.; Hopkinson, R.F.; Milewska, E.; Papadopol, P. Development and Testing of Canada-Wide Interpolated Spatial Models of Daily Minimum–Maximum Temperature and Precipitation for 1961–2003. *J. Appl. Meteorol. Clim.* **2009**, *48*, 725–741. [CrossRef]

51. Mpelasoka, F.S.; Chiew, F.H.S. Influence of Rainfall Scenario Construction Methods on Runoff Projections. *J. Hydrometeorol.* **2009**, *10*, 1168–1183. [CrossRef]

52. Sulis, M.; Paniconi, C.; Marrocu, M.; Huard, D.; Chaumont, D. Hydrologic response to multimodel climate output using a physically based model of groundwater/surface water interactions. *Water Resour. Res.* **2012**, *48*, W12510. [CrossRef]

53. Grinevskiy, S.; Pozdniakov, S.; Dedulina, E. Regional-Scale Model Analysis of Climate Changes Impact on the Water Budget of the Critical Zone and Groundwater Recharge in the European Part of Russia. *Water* **2021**, *13*, 428. [CrossRef]

54. Luoma, S.; Okkonen, J. Impacts of Future Climate Change and Baltic Sea Level Rise on Groundwater Recharge, Groundwater Levels, and Surface Leakage in the Hanko Aquifer in Southern Finland. *Water* **2014**, *6*, 3671–3700. [CrossRef]

55. Rathay, S.; Allen, D.; Kirste, D. Response of a fractured bedrock aquifer to recharge from heavy rainfall events. *J. Hydrol.* **2018**, *561*, 1048–1062. [CrossRef]

56. Sulis, M.; Paniconi, C.; Rivard, C.; Harvey, R.; Chaumont, D. Assessment of climate change impacts at the catchment scale with a detailed hydrological model of surface-subsurface interactions and comparison with a land surface model. *Water Resour. Res.* **2011**, *47*, W01513. [CrossRef]

57. Prein, A.F.; Rasmussen, R.M.; Ikeda, K.; Liu, C.; Clark, M.P.; Holland, G.J. The future intensification of hourly precipitation extremes. *Nat. Clim. Chang.* **2016**, *7*, 48–52. [CrossRef]

58. Nemri, S.; Kinnard, C. Comparing calibration strategies of a conceptual snow hydrology model and their impact on model performance and parameter identifiability. *J. Hydrol.* **2019**, *582*, 124474. [CrossRef]

59. Melsen, L.A.; Guse, B. Climate change impacts model parameter sensitivity—Implications for calibration strategy and model diagnostic evaluation. *Hydrol. Earth Syst. Sci.* **2021**, *25*, 1307–1332. [CrossRef]

60. Jaramillo, F.; Prieto, C.; Lyon, S.W.; Destouni, G. Multimethod assessment of evapotranspiration shifts due to non-irrigated agricultural development in Sweden. *J. Hydrol.* **2013**, *484*, 55–62. [CrossRef]

61. Destouni, G.; Jaramillo, F.; Prieto, C. Hydroclimatic shifts driven by human water use for food and energy production. *Nat. Clim. Chang.* **2012**, *3*, 213–217. [CrossRef]

62. Guerrero-Morales, J.; Fonseca, C.; Goméz-Albores, M.; Sampedro-Rosas, M.; Silva-Gómez, S. Proportional Variation of Potential Groundwater Recharge as a Result of Climate Change and Land-Use: A Study Case in Mexico. *Land* **2020**, *9*, 364. [CrossRef]

63. Verburg, P.H.; Neumann, K.; Nol, L. Challenges in using land use and land cover data for global change studies. *Glob. Chang. Biol.* **2011**, *17*, 974–989. [CrossRef]

64. Koirala, S.; Jung, M.; Reichstein, M.; De Graaf, I.; Camps-Valls, G.; Ichii, K.; Papale, D.; Ráduly, B.; Schwalm, C.R.; Tramontana, G.; et al. Global distribution of groundwater-vegetation spatial covariation. *Geophys. Res. Lett.* **2017**, *44*, 4134–4142. [CrossRef]

65. Groupe Agéco. *Recherche Participative D'alternatives Durables Pour La Gestion De L'eau En Milieu Agricole Dans Un Contexte De Changement Climatique (RADEAU1) (Participative Research for Sustainable Options in Water Management in Agricole Region and within a Climate Change Context)*; Ministère de l'agriculture, des pêcheries et de l'alimentation, Fonds Vert: Quebec City, QC, Canada, 2019; p. 332. Available online: https://www.agrireseau.net/documents/Document_101346.pdf (accessed on 14 November 2021).

Article

Hydro-Meteorological Trends in an Austrian Low-Mountain Catchment

Gerald Krebs [1,2,*], David Camhy [1] and Dirk Muschalla [1]

[1]   Institute of Urban Water Management and Landscape Water Engineering, Graz University of Technology, 8010 Graz, Austria; david.camhy@tugraz.at (D.C.); d.muschalla@tugraz.at (D.M.)

[2]   Institute of Hydraulic Engineering and Water Resources Management, Graz University of Technology, 8010 Graz, Austria

*   Correspondence: gerald.krebs@tugraz.at; Tel.: +43-316-873-8159

Citation: Krebs, G.; Camhy, D.; Muschalla, D. Hydro-Meteorological Trends in an Austrian Low-Mountain Catchment. *Climate* **2021**, *9*, 122. https://doi.org/10.3390/cli9080122

Academic Editor: Mohammad Valipour

Received: 2 June 2021
Accepted: 26 July 2021
Published: 29 July 2021

Publisher's Note: MDPI stays neutral with regard to jurisdictional claims in published maps and institutional affiliations.

Abstract: While ongoing climate change is well documented, the impacts exhibit a substantial variability, both in direction and magnitude, visible even at regional and local scales. However, the knowledge of regional impacts is crucial for the design of mitigation and adaptation measures, particularly when changes in the hydrological cycle are concerned. In this paper, we present hydro-meteorological trends based on observations from a hydrological research basin in Eastern Austria between 1979 and 2019. The analyzed variables include air temperature, precipitation, and catchment runoff. Additionally, the number of wet days, trends for catchment evapotranspiration, and computed potential evapotranspiration were derived. Long-term trends were computed using a non-parametric Mann–Kendall test. The analysis shows that while mean annual temperatures were decreasing and annual temperature minima remained constant, annual maxima were rising. Long-term trends indicate a shift of precipitation to the summer, with minor variations observed for the remaining seasons and at an annual scale. Observed precipitation intensities mainly increased in spring and summer between 1979 and 2019. Catchment actual evapotranspiration, computed based on catchment precipitation and outflow, showed no significant trend for the observed time period, while potential evapotranspiration rates based on remote sensing data increased between 1981 and 2019.

Keywords: hydrological research basin; precipitation; temperature; long-term trends; climate change; evapotranspiration

1. Introduction

It is well documented that the climate is changing [1–3]. Impacts are seen as globally rising temperatures [2,4] with a reduced number of cold days and nights and an increased number of warm days and nights [4], an altered depth [5–7] and duration of snow and ice cover [6–8], changing precipitation [4,9–11] and river flow regimes [12–14], or an increased number of extreme events [2,4,15]. However, the magnitude and impact direction of major climate variables, such as temperature, precipitation, catchment runoff, and evapotranspiration in both climate observations and projections vary significantly at the global and regional scale [16,17].

While there is a consensus on global warming [2] supported by many studies (e.g., [15,18,19]), some areas experienced decreasing mean, maximum, or minimum temperatures 1951-2002 [20]. Precipitation observations indicate minor global changes despite a large, compensating variability with a decrease observed in the subtropics and Southern Europe [21,22], Southern Asia and Africa and increases observed in North America, South America, Eurasia, and North and Central Europe [11,22–24]. Furthermore, a seasonal shift of precipitation has been reported (e.g., [19,25]).

With respect to catchment runoff, a decrease was observed for some basins in China [18,26], while an increase in runoff was reported for other Chinese basins [27] or North-Eastern USA [28]. Blaschke et al. [24] report only minor changes in runoff for

Austrian catchments in the past 50 years but predict a future runoff reduction for summer and an increase for winter. As for precipitation, a seasonal runoff shift has been reported (e.g., [13]). Furthermore, an increase in flooding events, particularly in Alpine areas, has been observed [14].

Several studies report increasing potential evapotranspiration trends for most of the Northern hemisphere (e.g., [19,29–33]), while China experienced decreasing evapotranspiration rates over the past 50 years [34]. Some of these studies confirm the trend that dry areas become drier and wet areas become wetter, while some contradict [23,35].

The validation of observations is one of the most important tasks during hydrological assessments as faulty data obviously provoke wrong analysis results and conclusions. At the same time, particularly the validation of precipitation measurements is very demanding due to the spatial and temporal variability of rainfall and its stochastic nature. An appropriate validation strategy depends on several factors, such as the spatial distribution of stations, the recording and analysis frequency or the type of measurement device. While there is no standardized procedure that is generally applicable, validation strategies commonly comprise the following steps: (i) identification of documented defects, (ii) device-specific boundaries, (iii) climatological boundaries, (iv) temporal variability, (v) intra-stational validation, and (vi) inter-stational variability [36,37].

The literature shows that the impact of climate change is widely acknowledged. At the same time, it is obvious that the impacts greatly vary at a regional and even local scale. However, this knowledge is crucial to develop measures to mitigate and counteract hydrological climate change impacts at the regional and local scale. Furthermore, we aim to investigate whether large-scale climate observations or projections also hold for smaller catchments where hydro-meteorological conditions may be very site-specific. For this purpose, we analyzed the hydro-meteorological data from a hydrological research catchment in Styria, Eastern Austria, which has been monitored since 1979. Analyzed climate variables include precipitation depth and intensities, number of wet days, air temperature, river flow, and actual and potential evapotranspiration.

2. Materials and Methods

2.1. Hydrological Research Catchment Pöllau

The hydrological research basin (HRB) *Pöllau* was established in 1978 [38,39] and is currently operated by the Institute of Urban Water Management and Landscape Water Engineering at Graz University of Technology in cooperation with the Department 14 of the Federal State Styria. The decision to establish an HRB in the *Pöllau* sub-basin was based on a number of reasons: (i) the confining arched mountain ridge allows a clear delineation of the catchment, (ii) the loamy soils are characterized by low storage capacities, minimizing the influence of subsurface flow on catchment hydrology, and (iii) the climate of the catchment with heavy storm events in the summer and relatively dry winters is representative for the Eastern alpine foothills [40]. The catchment covers 58.3 km^2 and is located in Styria, Austria, about 60 km northeast of the city of Graz (Figure 1). The elevation of the catchment ranges from 398 to 1279 m, and the catchment land-cover is dominated by forest (ca. 43.8%) and grass- and cropland (ca. 51.8%) with a low degree of discontinuous urban fabric (ca. 4.4%) [41].

The catchment comprises two main sub-catchments that are monitored: (i) the sub-catchment Saifenbach/Dürre Saifen covering 23 km^2 (monitored 1997–2005 and since 2018) and (ii) the sub-catchment Prätisbach covering 21 km^2 (monitored since 1980). Additionally, the discharge at the joint catchment outlet of both the sub-catchments has been monitored since 1980. Characteristic catchment properties are given in Table 1.

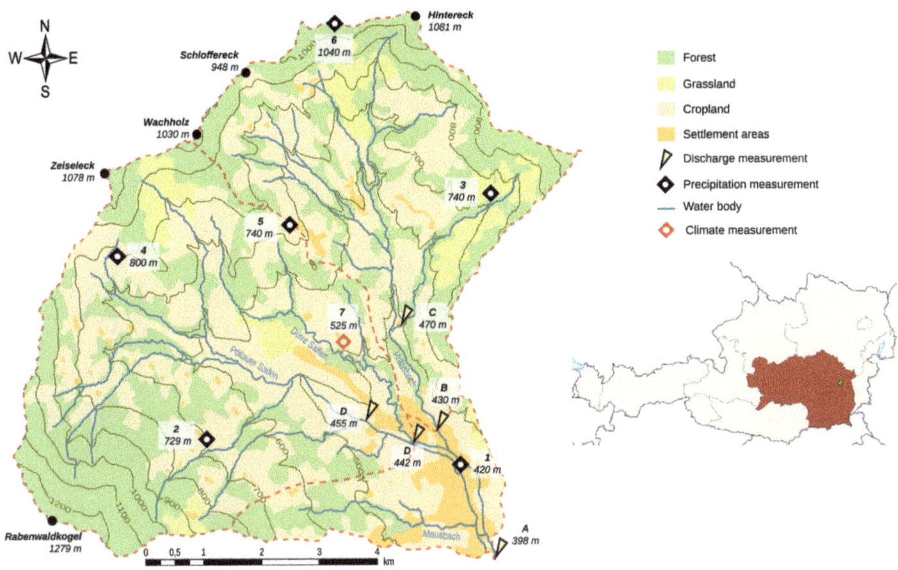

Figure 1. Overview of the catchment *Pöllau* (discharge measurement *A*) with the sub-catchment *Prätisbach* (discharge measurement *B*) in the west and the sub-catchment *Dürre Saifen* (discharge measurement *D*) in the east, and the locations of the precipitation measurements.

Table 1. Overview of the catchment properties [40].

Area	58.3 km^2
Land-use	forest 43.8%, grass- and cropland 51.8%, settlement 4.4%
Stream density	1.87 km km^{-2} or 0.0019 m m^{-2}
Geology	Crystalline basement rock 82.7%, tertiary hill country 12.7%, quaternary deposits 4.3%
Elevation range	398–1279 m.a.s.l
Discharge characteristics	Q_{min} 0.04 m^{3}s^{-1}; Q_{max} 92.14 m^{3}s^{-1}; Q_{mean} 0.49 m^{3}s^{-1}; Mean runoff coefficient 0.31 (1979–2004)

2.2. Data

The first precipitation measurement gauge in the HRB *Pöllau* was installed in 1979 (*1*, see Figure 1 and Table 2). During the following year (1980), an additional five precipitation gauges were installed, and two stream gauges (the catchment outlet *A* and the sub-catchment *B*) were constructed and taken into operation. The precipitation monitoring at the meteorological station (*7*) started in 1982, whereas the observation of climate variables started in 1991. The stream gauge *C* started operation in 1988 but was destroyed during a massive flood in 1997. The gauge was then reconstructed in 2000, but after further flood damage in 2007, the gauge was not put back into operation. The stream gauge *D* was constructed in 1997, but due to the challenging measurement location, monitoring was abandoned in 2005. The gauge was reconstructed 500 m upstream in 2018 and is, together with the gauges *A* and *B*, currently operating.

The currently operated precipitation gauges are rather symmetrically distributed over the catchment area and located at elevations between 420 and 1040 m.a.s.l. Initially, all 7 precipitation gauges were tipping buckets with a resolution of 0.1 mm. Since the year 2011, 6 stations have been equipped with rain scales (type Ott *Pluvio*^{2} [42]) operated at a 1 min recording interval. The rain scales in the catchment are not equipped with heaters, which hampers the record of snowfall. Due to this shortcoming, only snow that stayed on the gauges and thereafter melted was recorded. The currently operated stream gauges monitor the entire catchment outflow (*A*) and the two main sub-catchments (Figure 1). The stream gauges are equipped with pressure sensors, calibrated with rating curves, and record at a 10–15 min interval.

Table 2. Stations, altitude (m.a.s.l), measured variables: *WL* (water level), *WT* (water temperature), *P* (precipitation), *T* (air temperature), *p* (air pressure), *rH* (relative humidity), *Ra* (solar radiation), *ST* (soil temperature), *SM* (soil moisture), *WS* (wind speed), *WD* (wind direction), and data availability.

Station	Altitude	Observed Variables	Data Availability
A	398	*WL, WT*	1980–
B	415	*WL, WT*	1980–
C	418	*WL, WT*	1988–1997, 2000–2007
D	455	*WL, WT*	1997–2005, 2018–
1	424	*P*	1979–
2	729	*P*	1980–
3	740	*P*	1980–
4	800	*P*	1980–
5	740	*P*	1980–
6	1040	*P*	1980–
7	525	*P, T, p, rH, Ra, ST, SM, WS, WD*	1980–

Land-cover changes due to urbanization, agriculture or forestation can, along with potential climate change, significantly affect the catchment water balance, hampering the attribution of observed long-term changes to a single driver. Therefore the catchment land-use was analyzed based on the CORINE land-cover datasets available for the years 1990, 2000, 2006, 2012, and 2018 [43]. Between 1990 and 2018, forested areas decreased by 0.27 km^2 and were mostly replaced by settlements growing by 0.3 km^2, while the fraction of agricultural landuse remained almost constant with a decrease of 0.03 km^2 (Table 3).

Table 3. Land-cover in the catchment *Pöllau* in 1990, 2000, 2006, 2012, and 2018 based on the CORINE land-cover datasets [43].

Land-Cover		1990	2000	2006	2012	2018	Δ1990–2018
Discontinuous urban fabric		2.21	2.48	2.48	2.51	2.51	0.30
Mixed forest	(km^2)	9.34	9.19	9.19	9.19	9.19	−0.15
Coniferous forest		16.30	16.16	16.16	16.18	16.18	−0.12
Agricultural areas and pastures		30.05	30.07	30.07	30.02	30.02	−0.03
Discontinuous urban fabric		3.8	4.3	4.3	4.4	4.4	0.6
Mixed forest	(%)	16.1	15.9	15.9	15.9	15.9	−0.2
Coniferous forest		28.2	28.0	28.0	28.0	28.0	−0.2
Agricultural areas and pastures		51.9	51.9	51.9	51.8	51.8	−0.1

2.3. Data Validation

To exclude as much doubtful data as possible from the subsequent analysis, the available measurements were first validated on a daily basis according to the following procedure: (i) identification of documented defects, (ii) device-specific boundaries, (iii) climatological boundaries, (iv) temporal variability, (v) intra-stational validation, and (vi) inter-stational variability [36,37].

The validation steps (i)–(vi) were applied for rainfall and discharge observations. The comparison of daily precipitation observations after validation shows a good correlation (Pearson correlation 0.91), allowing the conclusion that the seven stations mostly recorded similar values (Figure 2). Discharge measurements were validated using cumulative sums of available gauges. An inter-stational validation for temperature data was not directly possible, as this variable is recorded at only one location within the catchment. However, the general observed pattern was compared with regionally available temperature observations for consistency.

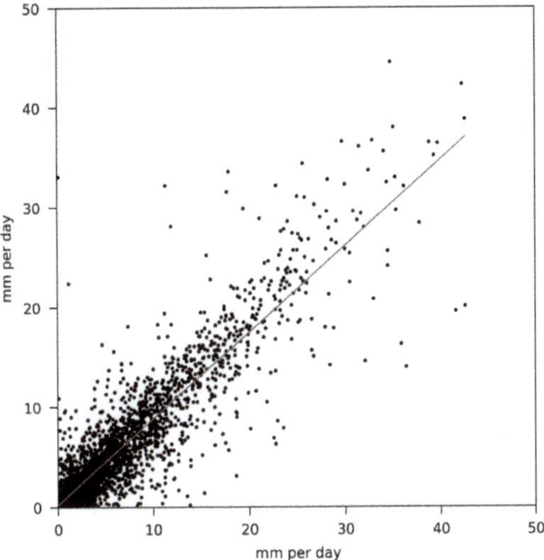

Figure 2. Scatter of daily recordings of each station against each station (Pearson correlation 0.91).

2.4. Data Analysis

Long-term hydrological trends and their significance were computed using the non-parametric modified Mann–Kendall test [44] to reduce the influence of serial correlation. Additionally, the Theil–Sen robust estimate was computed [45,46] to evaluate the magnitude of the trend. This approach has been successfully used to assess climate developments in numerous earlier studies (e.g., [47–50]) and was therefore applied in the current study.

Long-term trends of air temperatures were analyzed based on mean annual temperatures, on the one hand, and on mean seasonal temperatures recorded at the climate station 7 on the other hand. The seasons were defined as spring (March, April, May), summer (June, July, August), autumn (September, October, November) and winter (December, January, February). Seasonal trends were computed as annual trends that might be balanced by seasonal changes.

The conducted precipitation analyses comprised long-term trends of annual and seasonal (seasons as defined above), precipitation depths, and long-term trends of precipitation intensities for different durations (60, 120, 240 min). Precipitation depth was analyzed as the catchment mean sum (mean of the station recordings that fulfilled the validation criteria). Additionally, the trend of the number of annual wet days in the catchment was analyzed.

Long-term trends for the catchment discharge were analyzed for gauge *A*, while the remaining gauges were utilized for data validation only.

As for precipitation and temperature, long-term flow trends were also analyzed at a seasonal scale to identify temporal shifts in stream flow behavior.

The catchment water balance was computed based on observed precipitation runoff to assess the long-term development of actual evapotranspiration in the catchment. The computation includes a number of simplifications: (i) groundwater outflow of the catchment is not considered (no data available), (ii) land-cover changes were minor in the catchment during the observation period and therefore not further considered, and (iii) only years are taken into account, where available data allow for computation of annual runoff values. The simplifications yield the following water balance:

$$ET = P - R \tag{1}$$

where ET is actual evapotranspiration (mm), P is observed catchment precipitation (mm), and R is observed catchment runoff (mm). Additionally, potential catchment evapotranspiration (PET) was computed based on remote sensing data from the Copernicus Climate Data Store [51] as global radiation observations at the catchment climate station were insufficiently complete to allow for PET computation. PET was computed according to de Bruin et al. [52] using temperature, surface solar radiation and top of the atmosphere incident solar radiation. This approach was selected as wind speed data recordings were incomplete at the catchment climate station. Wind speed substantially influences PET, and thus, this simplification may underestimate computed PET rates.

3. Results

3.1. Temperature Trends

The mean annual air temperature at the climate station 7 between 1991 and 2019 is 9.8 °C, with the maximum annual mean recorded in 1995 (11.5 °C) and the minimum annual mean recorded in 1991 (7.6 °C). The long-term development of the mean annual temperature shows a negative trend with decreasing annual mean air temperature recordings (Figure 3).

While the development of annual minima shows no significant trend, annual temperature maxima were increasing between 1991 and 2019. The mean annual minimum 1991–2019 is −13.4 °C, with the lowest recording in 2009 (−19.2 °C) and the highest recording in 2015 (−8.3 °C). The mean annual maximum 1991–2019 is 32.5 °C, with the lowest recording in 1997 (29.1 °C) and the highest recording in 2003 and 2016 (37.4 °C) (Figure 3).

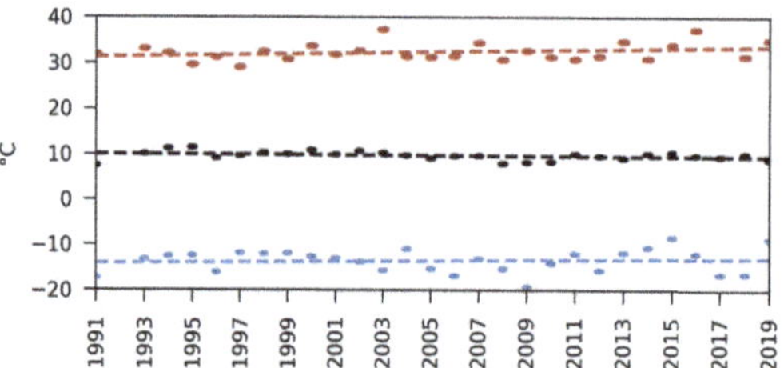

Figure 3. Annual mean (black), maxima (red) and minima (blue) of the air temperature and their trends 1979–2019.

The mean winter temperature between 1991 and 2019 is 0.3 °C shows a decreasing trend, with the lowest value recorded in 2009 (−3.2 °C) and the highest recording in 1994 (2.7 °C). Both winter minima (mean of −13.4 °C, with the lowest recording in 2009 (−19.2 °C) and the highest recording in 2015 (−8.3 °C)) and maxima 1991–2019 (mean of 16.5 °C, with the lowest recording in 1996 (10.6 °C) and the highest recording in 2011 (20.3 °C)) show no significant trend (Figure 4 top left).

The mean spring temperature between 1991 and 2019 is 10.0 °C and shows a similarly decreasing trend as observed for winter. The lowest mean was recorded in 2009 (7.6 °C) and the highest recording in 1995 (12.1 °C). The trend of spring minima is decreasing around a mean minimum of −5.5 °C, with the lowest recording in 2018 (−16.3 °C) and the highest record observed in 2011 (−1.2 °C). The trend of spring maxima 1991–2019 is also decreasing around 27.0 °C, with the lowest recording in 1991 (23.1 °C) and the highest recording in 1999 (30.9 °C) (Figure 4 top right).

The mean temperature during summer between 1991 and 2019 shows no trend staying at 19.9 °C, with the lowest recording in 2008 (17.5 °C) and the highest recording in 2003 (22.2 °C). The trend of summer minima is also not significant at 6.3 °C, with the lowest recording in 2006 (2.7 °C) and the highest recording in 2019 (12.7 °C). The trend of summer maxima 1991–2019 is increasing around 35.5 °C, with the lowest recording in 1997 (29.1 °C) and the highest recording in 2003 and 2016 (37.4 °C) (Figure 4 bottom left).

The trend of the mean autumn temperature 1991–2019 is not significant around 9.8 °C, with the lowest recording in 2008 (6.0 °C) and the highest recording in 1995 (13.4 °C). The trend of autumn minima is decreasing around −4.9 °C, with the lowest recording in 2008 (−9.8 °C) and the highest recording in 1995 (1.0 °C). The trend of autumn maxima 1991–2019 is increasing around 25.7 °C, with the lowest recording in 2010 (22.7 °C) and the highest recording in 2015 (31.5 °C) (Figure 4 bottom right). A comprehensive summary of the observed temperature trends, including statistical trend properties, is given in Table 4.

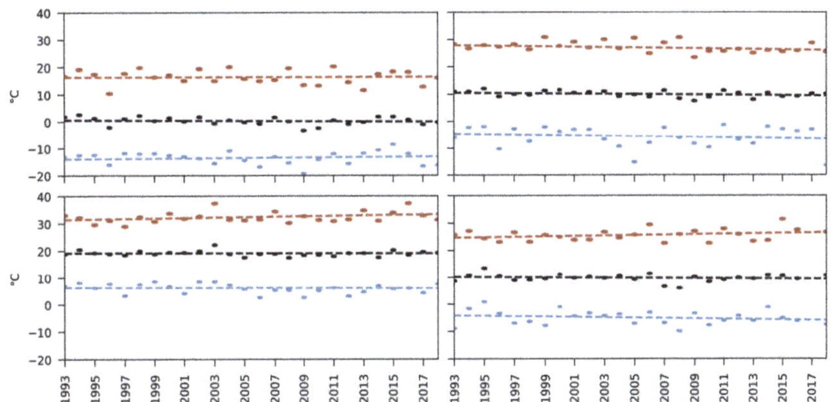

Figure 4. Mean (black), maxima (red) and minima (blue) of the air temperature and the trend 1979–2019 for the winter (**top left**), spring (**top right**), summer (**bottom left**), and autumn (**bottom right**).

Table 4. Summary of the climate variable trends for the catchment *Pöllau*.

Assessment Period	Variable	Unit	Y-W Trend	*p*-Value	T-S Slope
	mean air temperature	(°C)	decrease	1.4E-03	−3.1E-02
	minimum air temperature	(°C)	no trend	2.3E-01	2.7E-02
	maximum air temperature	(°C)	increase	1.2E-03	6.3E-02
	precipitation depth	(mm)	no trend	9.3E-02	6.0E-01
	annual wet days	(days)	no trend	7.1E-01	−7.7E-02
	precipitation intensity	(mm/60 min)	no trend	5.7E-01	4.0E-04
Annual	precipitation intensity	(mm/120 min)	no trend	4.3E-01	5.0E-03
	precipitation intensity	(mm/240 min)	no trend	5.9E-01	7.0E-03
	mean river flow	$(\text{m}^3\text{s}^{-1})$	decrease	4.2E-03	−2.4E-02
	minimum river flow	$(\text{m}^3\text{s}^{-1})$	increase	5.2E-03	1.0E-03
	maximum river flow	$(\text{m}^3\text{s}^{-1})$	no trend	7.3E-01	−1.1E-01
	actual evapotranspiration	(mm)	no trend	7.1E-01	1.3E-01
	potential evapotranspiration	(mm)	increase	0.0E00	1.4E00
	mean air temperature	(°C)	decrease	1.1E-02	−3.7E-02
	minimum air temperature	(°C)	no trend	2.2E-01	−1.5E-02
	maximum air temperature	(°C)	no trend	6.5E-01	2.8E-02
	precipitation depth	(mm)	no trend	4.5E-01	−1.5E-01
	precipitation intensity	(mm/60 min)	increase	5.5E-04	1.7E-02
Winter	precipitation intensity	(mm/120 min)	no trend	5.8E-02	9.0E-03
	precipitation intensity	(mm/240 min)	no trend	1.8E-01	1.4E-02
	mean river flow	$(\text{m}^3\text{s}^{-1})$	decrease	3.2E-03	−5.0E-03
	minimum river flow	$(\text{m}^3\text{s}^{-1})$	no trend	4.6E-01	3.0E-04
	maximum river flow	$(\text{m}^3\text{s}^{-1})$	no trend	1.8E-01	−9.0E-03

Table 4. *Cont.*

Assessment Period	Variable	Unit	Y-W Trend	*p*-Value	T-S Slope
Spring	mean air temperature	(°C)	decrease	2.6E-05	−4.7E-02
	minimum air temperature	(°C)	decrease	1.1E-02	−4.6E-02
	maximum air temperature	(°C)	decrease	3.0E-04	−9.9E-02
	precipitation depth	(mm)	no trend	3.4E-01	2.8E-01
	precipitation intensity	(mm/60 min)	increase	3.1E-04	1.1E-02
	precipitation intensity	(mm/120 min)	increase	7.1E-03	2.5E-02
	precipitation intensity	(mm/240 min)	increase	1.0E-02	4.9E-02
	mean river flow	(m^3s^{-1})	decrease	1.1E-03	−9.0E-03
	minimum river flow	(m^3s^{-1})	no trend	2.4E-01	-9.0E-04
	maximum river flow	(m^3s^{-1})	increase	3.0E-02	3.6E-02
Summer	mean air temperature	(°C)	no trend	7.9E-01	−1.0E-03
	minimum air temperature	(°C)	no trend	2.3E-01	−3.4E-02
	maximum air temperature	(°C)	increase	7.5E-05	6.3E-02
	precipitation depth	(mm)	increase	2.5E-06	2.1E00
	precipitation intensity	(mm/60 min)	no trend	1.3E-01	3.4E-03
	precipitation intensity	(mm/120 min)	increase	0.0E00	5.2E-02
	precipitation intensity	(mm/240 min)	increase	1.6E-05	5.4E-02
	mean river flow	(m^3s^{-1})	no trend	2.0E-01	−1.1E-02
	minimum river flow	(m^3s^{-1})	no trend	1.5E-01	−1.5E-03
	maximum river flow	(m^3s^{-1})	increase	4.4E-02	3.6E-01
Autumn	mean air temperature	(°C)	no trend	8.5E-01	−5.0E-03
	minimum air temperature	(°C)	decrease	6.2E-03	−6.9E-02
	maximum air temperature	(°C)	increase	1.0E-03	6.4E-02
	precipitation depth	(mm)	no trend	5.0E-01	−2.8E-01
	precipitation intensity	(mm/60 min)	no trend	7.9E-01	−8.1E-17
	precipitation intensity	(mm/120 min)	no trend	2.1E-01	−5.0E-03
	precipitation intensity	(mm/240 min)	no trend	7.8E-01	−4.2E-04
	mean river flow	(m^3s^{-1})	no trend	1.6E-01	7.3E-03
	minimum river flow	(m^3s^{-1})	increase	8.0E-03	1.7E-03
	maximum river flow	(m^3s^{-1})	no trend	2.8E-01	2.9E-02

3.2. Precipitation Trends

3.2.1. Precipitation Depth

The mean annual precipitation shows no significant trend between 1979 and 2019 around 608.9 mm, with the maximum mean recorded in 2014 (807.2 mm) and the minimum recorded in 2001 (364.3 mm). The annual maximum at a single station was recorded in 1996 at 4 (829.2 mm) and the annual minimum in 2001 at 7 (340.4 mm) (Figure 5). It is to be noted that the circumstance in which the rain scales are not equipped with heaters allows for the recording of only melted snowfall. Thus, winter and partly spring precipitation trends depend both on snowfall and melting occurring. The same is true for derived precipitation intensity trends.

The seasonal precipitation 1979–2019 shows an increasing trend for summer (June, July, August), while no significant trend was detected for spring (March, April, May), autumn (September, October, November), and winter (December, January, February) 1979–2019 (Figure 6).

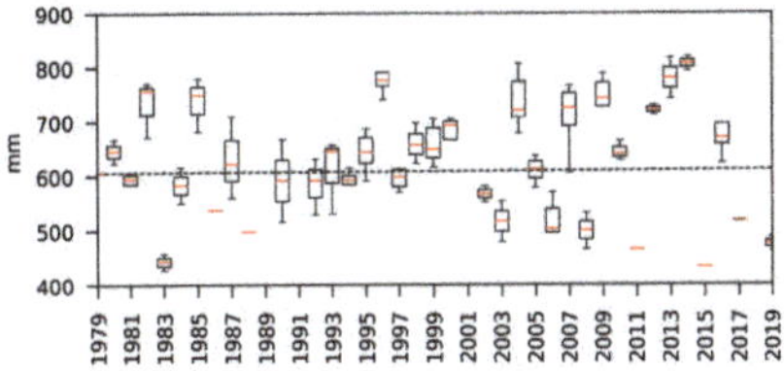

Figure 5. Annual precipitation of the 7 stations (25% and 75% percentile, median (red)) and trend 1979–2019 (dashed line).

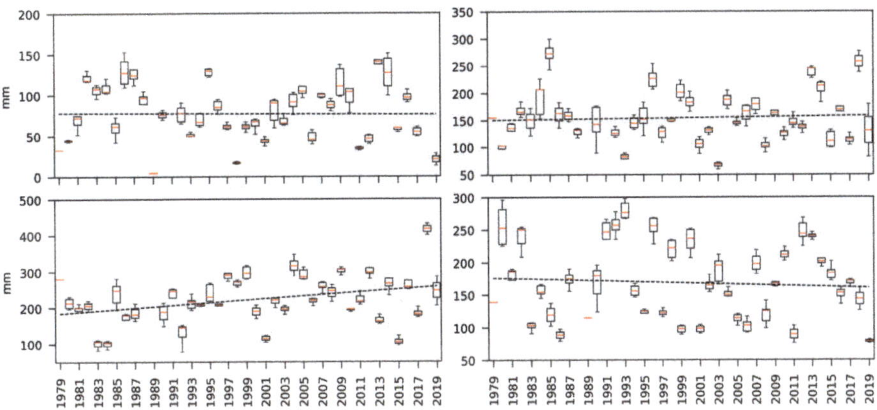

Figure 6. Precipitation of the 7 stations (25% and 75% percentile, median (red)) and trend 1979–2019 (dashed line) for winter (**top left**), spring (**top right**), summer (**bottom left**), and autumn (**bottom right**).

The mean winter precipitation in the catchment 1979–2019 was 73.3 mm, with the highest recording in 2013 (139.5 mm) and the lowest recording in 1998 (16.3 mm) (Figure 6 top left). The mean precipitation falling in the winter season accounted for 12% of the mean annual precipitation 1979–2019.

The mean spring precipitation accounted for 151.9 mm for 25% of the mean annual precipitation 1979–2019. The largest spring precipitation was recorded in 1985 (272.5 mm) and the smallest in 2003 (66.7 mm) (Figure 6 top right).

The mean summer precipitation shows a clearly increasing trend around 222.0 mm, accounting for 36% of the mean annual precipitation 1979–2019. The largest summer precipitation was recorded in 2018 (416.4 mm), and the smallest value was recorded in 1984 (99.1 mm) (Figure 6 bottom left).

The mean autumn precipitation 1979–2019 was around 168.2 mm, accounting for 27% of the mean annual precipitation. The largest autumn precipitation was recorded in 1993 (273.5 mm) and the smallest precipitation in 2019 (78.7 mm) (Figure 6 bottom right).

3.2.2. Wet Days

The number of wet days in the catchment remained constant, with a mean of 78 wet days per year (Figure 7). The highest number of wet days was recorded in 1979 with 105, while the smallest number of rainfall days was recorded in 2019 with 54.

3.2.3. Precipitation Intensities

Precipitation intensities for a duration of 60 min showed no significant trend at an annual level as well as for summer and autumn. However, an increasing trend was detected for winter and spring 1979–2019. Annual intensities for a duration of 120 min showed no significant trend as well as for winter and autumn, while spring and summer experienced increasing intensities (Figure 8). The trend for a longer duration of 240 min was not significant for winter and autumn as well as annually. However, as for the duration of 120 min, intensities were increasing for spring and summer. A comprehensive summary of the observed precipitation trends, including statistical trend properties, is given in Table 4.

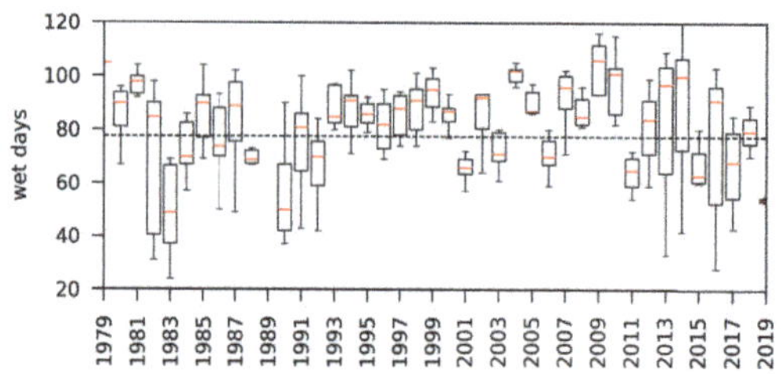

Figure 7. Annual wet days recorded at the 7 stations (25% and 75% percentile, median (red)) and trend 1979–2019 (dashed line).

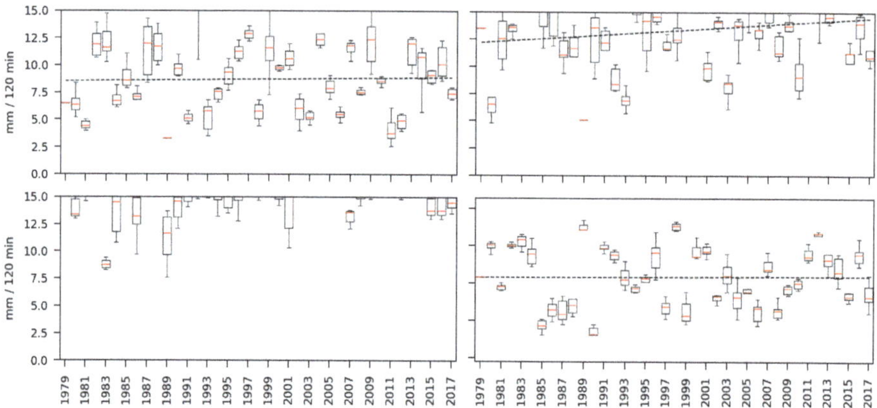

Figure 8. Seasonal maximum precipitation intensities for 120 minutes of the 7 stations (25% and 75% percentile, median (red)) and the trend 1979–2017 (dashed line).

3.3. River Flow Trends

The annual mean flow 1981–2016 at the catchment outlet A shows a decreasing trend around 1.10 m³s⁻¹, with the maximum mean flow observed in 1998 (3.01 m³s⁻¹) and the minimum mean flow observed in 2016 (0.12 m³s⁻¹) (Figure 9 left).

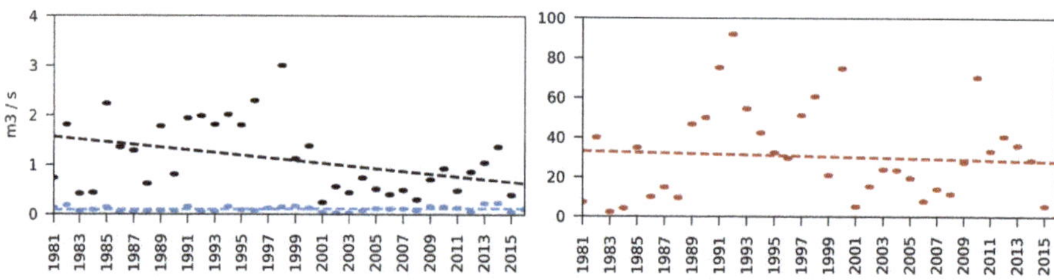

Figure 9. Annual mean (black, **left**), minimum (blue, **left**) and maximum (red, **right**) flow at Saifenbach and linear trends 1981–2016.

Observed mean annual minimum flows were increasing 1981–2016 around 0.11 m³s⁻¹ with the smallest recording in 2002 (0.03 m³s⁻¹) and the largest recording in 2014 (0.24 m³s⁻¹)

(Figure 9 left). Observed mean annual maximum flows showed no significant trend 1981–2016 around 31.10 m^{3}s^{-1}, with the largest observation in 1992 (92.14 m^{3}s^{-1}) and the smallest observation in 2015 (5.61 m^{3}s^{-1}) (Figure 9 right).

Mean winter flows 1981–2016 show, as already observed for annual flows, a decreasing trend around 0.51 m^{3}s^{-1}, with the lowest observation in the winter 2016 (0.12 m^{3}s^{-1}) and the largest observation in the winter 1992 (1.99 m^{3}s^{-1}). Minimum winter flows showed no significant trend 1981–2016 around 0.14 m^{3}s^{-1}, with the lowest flow in 2002 (0.03 m^{3}s^{-1}) and the largest minimum in 2014 (0.35 m^{3}s^{-1}). Maximum winter flows also remained constant 1981–2016 at 4.11 m^{3}s^{-1}, with the highest flow recorded in 1992 (34.28 m^{3}s^{-1}) and the lowest maximum in 1984 (0.58 m^{3}s^{-1}) (Figure 10 top).

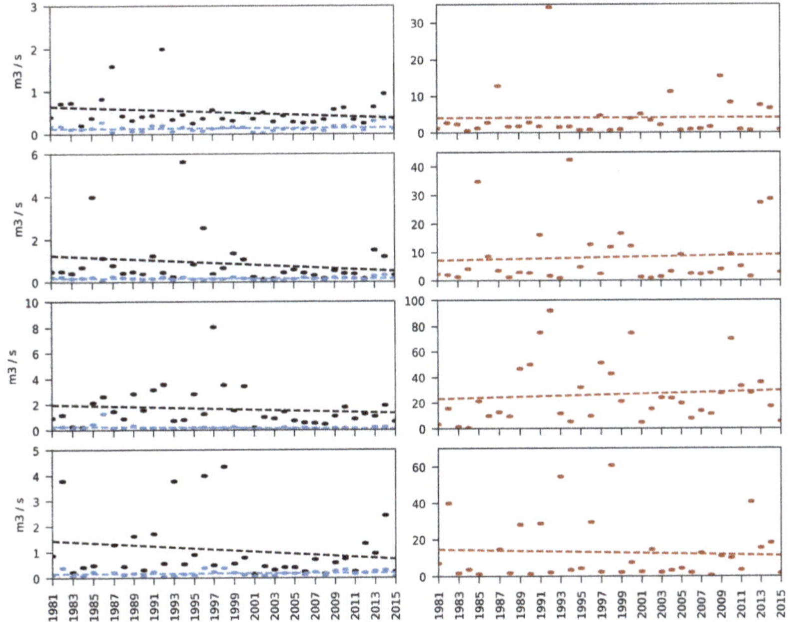

Figure 10. Mean (black), minimum (blue) and maximum (red) flow at Saifenbach and linear trends 1981–2016 for the winter (**top**), spring (2nd from top), summer (3rd from top), and autumn (**bottom**).

Mean spring flows 1981–2016 were decreasing around 0.88 m^{3}s^{-1}, with the lowest mean in 2002 (0.13 m^{3}s^{-1}) and the largest mean recorded in 1994 (5.62 m^{3}s^{-1}). Mean minimum spring flows show no trend at 0.17 m^{3}s^{-1}, with the lowest flow occurring in 2014 (0.33 m^{3}s^{-1}) and the highest minimum observed in 2002 (0.05 m^{3}s^{-1}). Mean maximum spring flows were increasing 1981–2016 around 8.27 m^{3}s^{-1}, with the largest recording in 1994 (42.42 m^{3}s^{-1}) and the smallest recording in 1993 (0.91 m^{3}s^{-1}) (Figure 10 2nd from top).

Mean summer flows 1981–2016 remained constant around 1.64 m^{3}s^{-1}, with the largest summer mean flow observed in 1997 (8.07 m^{3}s^{-1}) and the lowest mean in the summer 2001 (0.19 m^{3}s^{-1}). Summer minima show no trend 1981–2016 at 0.20 m^{3}s^{-1}, with the lowest observation in 2003 (0.04 m^{3}s^{-1}) and the highest in 1986 (1.30 m^{3}s^{-1}). Summer maxima increased 1981–2016 around 26.57 m^{3}s^{-1}, with the largest summer flow in 1992 (92.14 m^{3}s^{-1}) and the lowest maximum in 1984 (0.76 m^{3}s^{-1}) (Figure 10 3rd from top).

Mean autumn flows showed no trend 1981–2016 around 1.08 m^{3}s^{-1}, with the lowest mean recorded in 2001 (0.16 m^{3}s^{-1}) and the largest mean occurring in 1998 (4.35 m^{3}s^{-1}). Autumn minima decreased around 0.19 m^{3}s^{-1}, with the smallest flow recorded in autumn 1992 (0.06 m^{3}s^{-1}) and the largest minimum in 1982 (0.40 m^{3}s^{-1}). Maximum autumn flows remained constant 1981–2016 around 13.04 m^{3}s^{-1}, with the smallest maximum in autumn 2008 (0.73 m^{3}s^{-1}) and the largest autumn flow in 1998 (60.81 m^{3}s^{-1}) (Figure 10 bottom). A comprehensive summary of the observed runoff trends, including statistical trend properties, is given in Table 4.

3.4. Water Balance and Evapotranspiration

Particularly in the 1990s, flow measurements at *A* have large gaps preventing the computation of annual flow volumes. Thus, 21 years were available for the assessment of evapotranspiration based on precipitation and catchment runoff (Figure 11). The mean runoff fraction of the water balance 1981–2015 was 55% showing a decreasing trend. It is to be noted though that fewer data were available for the time period 1981–2000 (6 years) than for the period 2001–2015 (15 years). The highest runoff fraction was observed in 2014 with 89%, while the lowest fraction occurred in 2008 with only 30%. In absolute values, the catchment runoff ranged between 131 and 743 mm, with a mean of 338 mm per year.

Based on long-term precipitation and runoff trends, the actual evapotranspiration fraction showed no significant trend 1981–2015, with a mean of 45%, a minimum of 11% in 2014, and a maximum of 70% in 2008. In absolute numbers, actual catchment evapotranspiration 1981–2015 was around 257 mm, with a minimum of 92 mm in 2014 and a maximum of 451 mm in 2008.

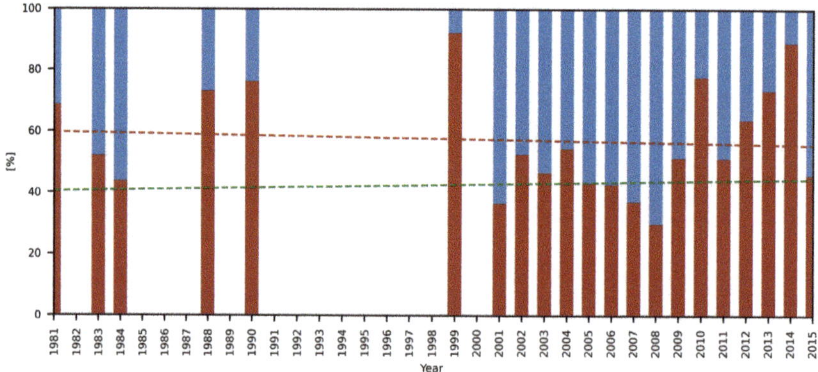

Figure 11. Annual water balance as fallen precipitation (100%, blue) and runoff fraction (red). The dashed lines mark the long-term trend of the runoff fraction (red) and the actual evapotranspiration fraction (green) 1981–2015. Missing years did not provide sufficient runoff data for a cumulative annual runoff value.

The potential catchment evapotranspiration (PET) was computed using remote sensing data for air temperature, global radiation, and top of the atmosphere solar radiation for the period 1981–2019. Catchment PET rates show an increasing trend 1981–2019 around a mean of 759 mm per year, with a minimum of 711 mm in 1995 and a maximum of 822 mm in 2002 (Figure 12).

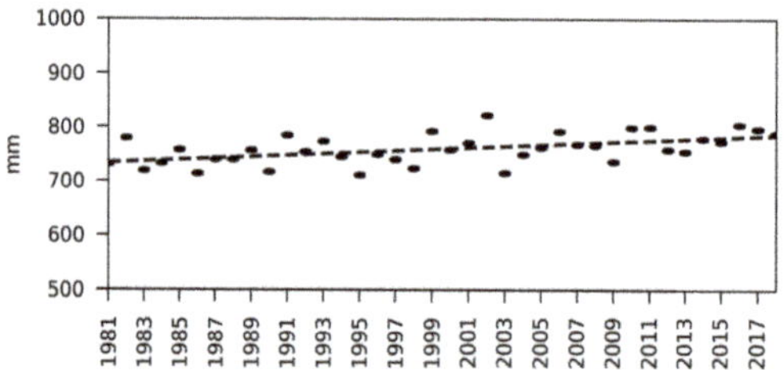

Figure 12. Annual potential evapotranspiration (PET) based on de Bruin [52] computed with remote sensing data [51] and trend 1981–2019 (dashed line).

4. Discussion

The mean annual air temperature in the catchment *Pöllau* has been decreasing since 1991, while annual minima remained constant and maxima were increasing. While the development of minima and maxima is a common consequence of ongoing climate change (e.g., [53,54]), the decreasing long-term development of the annual mean temperature in *Pöllau* is less often confirmed by literature (e.g., [20]) as clearly more often rising temperatures are reported (e.g., [15,18,19,55]). For Austria, rising temperatures have also been reported [56], which confirms that climate change impacts at the local or regional scale differ from large-scale assessments. On the other hand, the rising maximum temperatures in the catchment are in line with the APCC report [56]. It is to be noted that the observed time series in *Pöllau* covers approximately 30 years and is thus rather short for temperature change detection. It might, therefore, well be that the analyzed time period coincided with a period where warming in the catchment did not occur (see, e.g., [57]). This assumption is also confirmed by reports and studies addressing climate change in Austria (e.g., [56,58,59]).

The reported climate-change-induced perturbations to precipitation patterns are far more diverse than for air temperature. Increasing [11,23] and decreasing precipitation rates [20,21] were reported as well as areas where no change was detected [19,20,60]. Mean annual precipitation in *Pöllau* remained constant between 1979 and 2019. This observation is confirmed by the Austrian APCC report [56], reporting increased precipitation for the Austrian alpine areas and a decrease for South-East Austria since the beginning of observations. The *Pöllau* catchment falls in between these two areas in the Eastern alpine foothills. The seasonal precipitation analysis indicates a shift towards the summer season, for which an increasing trend was observed. The remaining seasons (spring, autumn, winter) showed no significant trend concerning the fallen precipitation 1979–2019. It is to be noted that the circumstance in which the rain scales are not equipped with heaters allows for the recording of only melted snowfall. Thus, winter and partly spring precipitation trends depend both on snowfall and melting occurring. Seasonal shifts in precipitation have also been reported by earlier studies (e.g., [9,10]), but it is to be noted that especially the climate change induced impact on precipitation shows obvious regional differences [56]. Precipitation intensities for the analyzed durations were increasing for spring and summer. While summer precipitation depth 1979–2019 was also increasing, it remained constant for spring, allowing the assumption of a reduction of events and at the same time, a higher event precipitation. For winter and autumn, no significant trends were detected, as already observed for the precipitation depth in these seasons. Furthermore, for the precipitation intensity trends, it is to be noted that snowfall was only recorded indirectly via melted snow.

Mean river flows at gauge *A* decreased annually as well as for spring and summer, while flow minima increased annually and for autumn, and flow maxima increased for spring and summer. These observations are in line with the APCC report [56]. At the same time, the precipitation depth increased only during the summer season and analyzed precipitation intensities during spring and summer. The rather opposite trends for the precipitation depth and mean river flows indicate that more water is evapotranspirated in the catchment during the warm season, and increasing flow maxima during spring and summer could be due to increasing precipitation intensities for the same seasons.

Based on observed catchment precipitation and runoff, the actual annual catchment evapotranspiration showed no significant trend between 1981 and 2015. It is to be noted though that only river flow at the catchment outlet was used for computation as subsurface flow data were not available. PET rates show an increasing trend 1981–2019 for the catchment. It is to be noted though that these rates were computed based on remote sensing data as local ground climate data were insufficiently complete for PET computation. Furthermore, the selected PET computation approach [52] did not account for wind speed, due to missing data, and may therefore underestimate catchment PET. These observations are confirmed by several studies reporting similar evapotranspiration trends for the Northern hemisphere [19,29–32] as well as by the APCC report [56].

5. Conclusions

The presented analyses of hydro-meteorological variables observed in a hydrological research basin in Eastern Austria mostly confirm the results of earlier studies. At the same time, the results confirm the assumption that climate change impacts vary regionally, and large-scale assessments cannot account for site-specific conditions.

- The decreasing trend of long-term mean annual temperatures in the catchments shows that climate change impacts can vary at the regional scale.
- The observed precipitation trends are in line with large-scale assessments, including the study catchment. However, precipitation recordings during the cold season were hampered by missing rain scale heaters. For a full assessment of precipitation developments and especially seasonal changes, heated rain scales should be used.
- Climate data observations such as global radiation, relative humidity, wind speed or soil moisture are of substantial importance to assess the drivers for the change in climate variables. Thus, a comprehensive monitoring is required to assess not only if but also why climate variables are changing.
- The impact of increasing precipitation intensities is seen in larger river flow maxima during spring and summer.
- Actual catchment evapotranspiration (AET) remained constant, while potential catchment evapotranspiration (PET) increased 1981–2019. It is to be noted that AET was computed based on river runoff that was not fully available for a significant number of years, hampering the assessment.
- The analysis of hydro-meteorological variable trends can be supported by numerical modeling approaches to evaluate the variations in hydrological and meteorological processes in more detail. This numerical assessment is currently conducted for the catchment *Pöllau*.

Author Contributions: Conceptualization, G.K. and D.M.; data curation, G.K.; formal analysis, G.K.; funding acquisition, D.M.; investigation, G.K.; methodology, G.K. and D.C.; project administration, G.K. and D.M.; supervision, D.M.; validation, G.K., D.C. and D.M.; visualization, G.K.; writing—original draft, G.K.; writing—review and editing, G.K., D.C. and D.M. All authors have read and agreed to the published version of the manuscript.

Funding: This research received no external funding.

Institutional Review Board Statement: Not applicable.

Informed Consent Statement: Not applicable.

Data Availability Statement: Restrictions apply to the availability of these data. Data were obtained from the Institute of Urban Water Management, Graz University of Technology and Department 14, Federal State of Styria, and are available from the authors with the permission of the Institute of Urban Water Management, Graz University of Technology and Department 14, Federal State of Styria.

Acknowledgments: Roland Fuchs and Christian Rath for operation and maintenance of the test catchment; Department 14 Styria (Robert Schatzl and Josef Quinz) for the support for the test catchment operation. Open Access Funding by the Graz University of Technology.

Conflicts of Interest: The authors declare no conflict of interest.

References

1. IPCC. *Climate Change 2014: Synthesis Report. Contribution of Working Groups I, II and III to the Fifth Assessment Report of the Intergovernmental Panel on Climate Change*; Technical Report; IPCC: Geneva, Switzerland, 2014.
2. IPCC. *Global Warming of 1.5 °C. An IPCC Special Report on the Impacts of Global Warming of 1.5 °C above Pre-Industrial Levels and Related Global Greenhouse Gas Emission Pathways, in the Context of Strengthening the Global Response to the Threat of Climate Change, Sustainable Development, and Efforts to Eradicate Poverty*; OCLC: 1056192590; Technical Report; IPCC: Geneva, Switzerland, 2018.
3. IPCC. *Climate Change 2013: The Physical Science Basis. Contribution of Working Group I to the Fifth Assessment Report of the Intergovernmental Panel on Climate Change*; Reporter: IPCC; Technical Report; Cambridge University Press: Cambridge, UK; New York, NY, USA, 2013.
4. USGCRP. *Climate Science Special Report: Fourth National Climate Assessment, Volume I*; Technical Report; U.S. Global Change Research Program: Washington, DC, USA, 2017.
5. Steger, C.; Kotlarski, S.; Jonas, T.; Schär, C. Alpine snow cover in a changing climate: A regional climate model perspective. *Clim. Dyn.* **2013**, *41*, 735–754. [CrossRef]
6. Olefs, M.; Koch, R.; Schöner, W.; Marke, T. Changes in Snow Depth, Snow Cover Duration, and Potential Snowmaking Conditions in Austria, 1961–2020—A Model Based Approach. *Atmosphere* **2020**, *11*, 1330. [CrossRef]
7. Bender, E.; Lehning, M.; Fiddes, J. Changes in Climatology, Snow Cover, and Ground Temperatures at High Alpine Locations. *Front. Earth Sci.* **2020**, *8*, 100. [CrossRef]
8. Gardner, A.S.; Moholdt, G.; Cogley, J.G.; Wouters, B.; Arendt, A.A.; Wahr, J.; Berthier, E.; Hock, R.; Pfeffer, W.T.; Kaser, G.; et al. A Reconciled Estimate of Glacier Contributions to Sea Level Rise: 2003 to 2009. *Science* **2013**, *340*, 852–857. [CrossRef] [PubMed]
9. Brugnara, Y.; Brunetti, M.; Maugeri, M.; Nanni, T.; Simolo, C. High-resolution analysis of daily precipitation trends in the central Alps over the last century. *Int. J. Climatol.* **2012**, *32*, 1406–1422. [CrossRef]
10. Brunetti, M.; Maugeri, M.; Nanni, T.; Auer, I.; Böhm, R.; Schöner, W. Precipitation variability and changes in the greater Alpine region over the 1800–2003 period. *J. Geophys. Res. Atmos.* **2006**, *111*. [CrossRef]
11. Trenberth, K. Changes in precipitation with climate change. *Clim. Res.* **2011**, *47*, 123–138. [CrossRef]
12. Arnell, N.W.; Gosling, S.N. The impacts of climate change on river flow regimes at the global scale. *J. Hydrol.* **2013**, *486*, 351–364. [CrossRef]
13. Blöschl, G.; Hall, J.; Parajka, J.; Perdigão, R.A.P.; Merz, B.; Arheimer, B.; Aronica, G.T.; Bilibashi, A.; Bonacci, O.; Borga, M.; et al. Changing climate shifts timing of European floods. *Science* **2017**, *357*, 588–590. [CrossRef]
14. Blöschl, G.; Viglione, A.; Merz, R.; Parajka, J.; Salinas, J.L.; Schöner, W. Auswirkungen des Klimawandels auf Hochwasser und Niederwasser. *Österreichische Wasser und Abfallwirtschaft* **2011**, *63*, 21–30. [CrossRef]
15. Luterbacher, J.; Dietrich, D.; Xoplaki, E.; Grosjean, M.; Wanner, H. European Seasonal and Annual Temperature Variability, Trends, and Extremes Since 1500. *Science* **2004**, *303*, 1499–1503. [CrossRef] [PubMed]
16. Hu, Z.; Chen, X.; Chen, D.; Li, J.; Wang, S.; Zhou, Q.; Yin, G.; Guo, M. "Dry gets drier, wet gets wetter": A case study over the arid regions of central Asia. *Int. J. Climatol.* **2019**, *39*, 1072–1091. [CrossRef]
17. Latif, M. Uncertainty in climate change projections. *J. Geochem. Explor.* **2011**, *110*, 1–7. [CrossRef]
18. Li, L.J.; Zhang, L.; Wang, H.; Wang, J.; Yang, J.W.; Jiang, D.J.; Li, J.Y.; Qin, D.Y. Assessing the impact of climate variability and human activities on streamflow from the Wuding River basin in China. *Hydrol. Process.* **2007**, *21*, 3485–3491. [CrossRef]
19. Chaouche, K.; Neppel, L.; Dieulin, C.; Pujol, N.; Ladouche, B.; Martin, E.; Salas, D.; Caballero, Y. Analyses of precipitation, temperature and evapotranspiration in a French Mediterranean region in the context of climate change. *Comptes Rendus Geosci.* **2010**, *342*, 234–243. [CrossRef]
20. Girvetz, E.H.; Zganjar, C.; Raber, G.T.; Maurer, E.P.; Kareiva, P.; Lawler, J.J. Applied Climate-Change Analysis: The Climate Wizard Tool. *PLoS ONE* **2009**, *4*, e8320. [CrossRef]
21. Cannarozzo, M.; Noto, L.V.; Viola, F. Spatial distribution of rainfall trends in Sicily (1921–2000). *Phys. Chem. Earth Parts A B C* **2006**, *31*, 1201–1211. [CrossRef]
22. Blöschl, G.; Montanari, A. Climate change impacts—Throwing the dice? *Hydrol. Process.* **2009**. [CrossRef]
23. Famiglietti, J.S.; Rodell, M. Water in the Balance. *Science* **2013**, *340*, 1300–1301. [CrossRef]
24. Blaschke, A.P.; Merz, R.; Parajka, J.; Salinas, J.; Blöschl, G. Auswirkungen des Klimawandels auf das Wasserdargebot von Grund- und Oberflächenwasser. *Österreichische Wasser und Abfallwirtschaft* **2011**, *63*, 31–41. [CrossRef]
25. Kumar, P. Seasonal rain changes. *Nat. Clim. Chang.* **2013**, *3*, 783–784. [CrossRef]
26. Zhang, S.; Hua, D.; Meng, X.; Zhang, Y. Climate change and its driving effect on the runoff in the "Three-River Headwaters" region. *J. Geogr. Sci.* **2011**, *21*, 963. [CrossRef]
27. Chen, Z.; Chen, Y.; Li, B. Quantifying the effects of climate variability and human activities on runoff for Kaidu River Basin in arid region of northwest China. *Theor. Appl. Climatol.* **2013**, *111*, 537–545. [CrossRef]
28. Wang, H.; Stephenson, S.R. Quantifying the impacts of climate change and land use/cover change on runoff in the lower Connecticut River Basin. *Hydrol. Process.* **2018**, *32*, 1301–1312. [CrossRef]
29. McCabe, G.J.; Wolock, D.M. Temporal and spatial variability of the global water balance. *Clim. Chang.* **2013**, *120*, 375–387. [CrossRef]

30. Vicente-Serrano, S.M.; Azorin-Molina, C.; Sanchez-Lorenzo, A.; Revuelto, J.; López-Moreno, J.I.; González-Hidalgo, J.C.; Moran-Tejeda, E.; Espejo, F. Reference evapotranspiration variability and trends in Spain, 1961–2011. *Glob. Planet. Chang.* **2014**, *121*, 26–40. [CrossRef]

31. Elferchichi, A.; Giorgio, G.A.; Lamaddalena, N.; Ragosta, M.; Telesca, V. Variability of Temperature and Its Impact on Reference Evapotranspiration: The Test Case of the Apulia Region (Southern Italy). *Sustainability* **2017**, *9*, 2337. [CrossRef]

32. Maček, U.; Bezak, N.; Šraj, M. Reference evapotranspiration changes in Slovenia, Europe. *Agric. For. Meteorol.* **2018**, *260–261*, 183–192. [CrossRef]

33. Calanca, P.; Roesch, A.; Jasper, K.; Wild, M. Global Warming and the Summertime Evapotranspiration Regime of the Alpine Region. *Clim. Chang.* **2006**, *79*, 65–78. [CrossRef]

34. McVicar, T.R.; Roderick, M.L.; Donohue, R.J.; Li, L.T.; Van Niel, T.G.; Thomas, A.; Grieser, J.; Jhajharia, D.; Himri, Y.; Mahowald, N.M.; et al. Global review and synthesis of trends in observed terrestrial near-surface wind speeds: Implications for evaporation. *J. Hydrol.* **2012**, *416–417*, 182–205. [CrossRef]

35. Feng, H.; Zhang, M. Global land moisture trends: Drier in dry and wetter in wet over land. *Sci. Rep.* **2015**, *5*, 18018. [CrossRef]

36. Maier, R.; Krebs, G.; Pichler, M.; Muschalla, D.; Gruber, G. Spatial Rainfall Variability in Urban Environments—High-Density Precipitation Measurements on a City-Scale. *Water* **2020**, *12*, 1157. [CrossRef]

37. Kirchengast, G.; Kabas, T.; Leuprecht, A.; Bichler, C.; Truhetz, H. WegenerNet: A Pioneering High-Resolution Network for Monitoring Weather and Climate. *Bull. Am. Meteorol. Soc.* **2013**, *95*, 227–242. [CrossRef]

38. Bergmann, H. *Experimentelle Hydrologische Forschung in der Steiermark*; Technical Report; Das hydrologische Versuchsgebiet Pöllau: Graz, Austria, 1981.

39. Bergmann, H. A Hydrological Research Basin in Austria: Planning and Aims. In Proceedings of the International Symposium On Hydrological Research Basins, Bern, Switzerland, 21–23 September 1982; pp. 23–30.

40. Ruch, C.; Vasvári, V.; Harum, T. Hydrologisches Versuchsgebiet Pöllau—25 Jahre Beobachtung. *Beiträge zur Hydrogeologie* **2006**, *55*, 45–58.

41. Krebs, G.; Weidemann, S.; Fuchs, R.; Muschalla, D. Hydrologisches Versuchsgebiet Pöllau—Hydrometeorologische Langzeitbetrachtungen. *Wasserland Steiermark* **2017**, *1*, 22–26.

42. OTT Pluvio^2—Regenmesser/Niederschlagsmesser—OTT HYDROMET CH. Available online: https://www.ott.com/de-at/produkte/meteorologie-29/ott-pluvio2-1-regenmesser-niederschlagsmesser-1520/ (accessed on 27 July 2021).

43. European Environment Agency. © European Union, Copernicus Land Monitoring Service. Page, f.ex. in 2018: "© European Union, Copernicus Land Monitoring Service 2018, European Environment Agency (EEA)". 2021. Available online: https://cds.climate.copernicus.eu/ (accessed on 27 July 2021).

44. Yue, S.; Wang, C. The Mann-Kendall Test Modified by Effective Sample Size to Detect Trend in Serially Correlated Hydrological Series. *Water Resour. Manag.* **2004**, *18*, 201–218. [CrossRef]

45. Theil, H. A Rank-Invariant Method of Linear and Polynomial Regression Analysis. In *Henri Theil's Contributions to Economics and Econometrics: Econometric Theory and Methodology*; Advanced Studies in Theoretical and Applied Econometrics; Raj, B., Koerts, J., Eds.; Springer: Dordrecht, The Netherlands, 1992; pp. 345–381. [CrossRef]

46. Sen, P.K. Estimates of the Regression Coefficient Based on Kendall's Tau. *J. Am. Stat. Assoc.* **1968**, *63*, 1379–1389. [CrossRef]

47. Alemu, Z.A.; Dioha, M.O. Climate change and trend analysis of temperature: The case of Addis Ababa, Ethiopia. *Environ. Syst. Res.* **2020**, *9*, 27. [CrossRef]

48. Mahmood, R.; Jia, S.; Zhu, W. Analysis of climate variability, trends, and prediction in the most active parts of the Lake Chad basin, Africa. *Sci. Rep.* **2019**, *9*, 6317. [CrossRef] [PubMed]

49. Peng, S.; Gang, C.; Cao, Y.; Chen, Y. Assessment of climate change trends over the Loess Plateau in China from 1901 to 2100. *Int. J. Climatol.* **2018**, *38*, 2250–2264. [CrossRef]

50. Merabtene, T.; Siddique, M.; Shanableh, A. Assessment of Seasonal and Annual Rainfall Trends and Variability in Sharjah City, UAE. *Adv. Meteorol.* **2016**, *2016*, e6206238. [CrossRef]

51. Hersbach, H.; Bell, B.; Berrisford, P.; Biavati, G.; Horányi, A.; Muñoz Sabater, J.; Nicolas, J.; Peubey, C.; Radu, R.; Rozum, I.; et al. ERA5 Hourly Data on Single Levels from 1979 to Present. 2018. Available online: https://cds.climate.copernicus.eu/ (accessed on 27 July 2021).

52. De Bruin, H.A.R.; Trigo, I.F.; Bosveld, F.C.; Meirink, J.F. A Thermodynamically Based Model for Actual Evapotranspiration of an Extensive Grass Field Close to FAO Reference, Suitable for Remote Sensing Application. *J. Hydrometeorol.* **2016**, *17*, 1373–1382. [CrossRef]

53. Easterling, D.; Horton, B.; Jones, P.; Peterson, T.; Karl, T.; Parker, D.; Salinger, M.; Razuvayev, V.; Plummer, N.; Jamason, P.; Folland, C. Maximum and Minimum Temperature Trends for the Globe. *Science* **1997**, *277*, 364–367. [CrossRef]

54. Ghasemi, A.R. Changes and trends in maximum, minimum and mean temperature series in Iran. *Atmos. Sci. Lett.* **2015**, *16*, 366–372. [CrossRef]

55. Folland, C.K.; Karl, T.R.; Salinger, M.J. Observed climate variability and change. *Weather* **2002**, *57*, 269–278. [CrossRef]

56. Kromp-Kolb, H.; Nakicenovic, N.; Steininger, K.; Gobiet, A.; Formayer, H.; Köppl, A.; Prettenthaler, F.; Stötter, J.; Schneider, J. *Österreichischer Sachstandsbericht Klimawandel 2014*; Verlag der Österreichischen Akademie der Wissenschaften: Vienna, Austria, 2014.

57. Easterling, D.R.; Wehner, M.F. Is the climate warming or cooling? *Geophys. Res. Lett.* **2009**, *36*. [CrossRef]

58. Gobiet, A.; Kotlarski, S.; Beniston, M.; Heinrich, G.; Rajczak, J.; Stoffel, M. 21st century climate change in the European Alps—A review. *Sci. Total. Environ.* **2014**, *493*, 1138–1151. [CrossRef]
59. Brunetti, M.; Lentini, G.; Maugeri, M.; Nanni, T.; Auer, I.; Böhm, R.; Schöner, W. Climate variability and change in the Greater Alpine Region over the last two centuries based on multi-variable analysis. *Int. J. Climatol.* **2009**, *29*, 2197–2225. [CrossRef]
60. Yong, B.; Ren, L.; Hong, Y.; Gourley, J.J.; Chen, X.; Dong, J.; Wang, W.; Shen, Y.; Hardy, J. Spatial–Temporal Changes of Water Resources in a Typical Semiarid Basin of North China over the Past 50 Years and Assessment of Possible Natural and Socioeconomic Causes. *J. Hydrometeorol.* **2013**, *14*, 1009–1034. [CrossRef]

Article

Performance Evaluation and Comparison of Satellite-Derived Rainfall Datasets over the Ziway Lake Basin, Ethiopia

Aster Tesfaye Hordofa [1,*], **Olkeba Tolessa Leta** [2], **Tena Alamirew** [3], **Nafyad Serre Kawo** [4] **and Abebe Demissie Chukalla** [5]

[1] Africa Centre of Excellence for Water Management, Addis Ababa University, Addis Ababa 1176, Ethiopia
[2] Bureau of Watershed Management and Modeling, St. Johns River Water Management District, 4049 Reid Street, Palatka, FL 32177, USA; OLeta@sjrwmd.com
[3] Ethiopian Institute of Water Resources, Addis Ababa University, Addis Ababa 1176, Ethiopia; tena.a@wlrc-eth.org
[4] School of Natural Resources, University of Nebraska-Lincoln, Lincoln, NE 68583, USA; nkawo2@huskers.unl.edu
[5] The Department of Land and Water Management, IHE Delft Institute for Water Education, 2611 Delft, The Netherlands; a.chukalla@un-ihe.org
* Correspondence: aster.tesfaye@aau.edu.et

Citation: Hordofa, A.T.; Leta, O.T.; Alamirew, T.; Kawo, N.S.; Chukalla, A.D. Performance Evaluation and Comparison of Satellite-Derived Rainfall Datasets over the Ziway Lake Basin, Ethiopia. *Climate* **2021**, *9*, 113. https://doi.org/10.3390/cli9070113

Academic Editors: Mohammad Valipour and Sayed M. Bateni

Received: 14 May 2021
Accepted: 28 June 2021
Published: 8 July 2021

Publisher's Note: MDPI stays neutral with regard to jurisdictional claims in published maps and institutional affiliations.

Abstract: Consistent time series rainfall datasets are important in performing climate trend analyses and agro-hydrological modeling. However, temporally consistent ground-based and long-term observed rainfall data are usually lacking for such analyses, especially in mountainous and developing countries. In the absence of such data, satellite-derived rainfall products, such as the Climate Hazard Infrared Precipitations with Stations (CHIRPS) and Global Precipitation Measurement Integrated Multi-SatellitE Retrieval (GPM-IMERG) can be used. However, as their performance varies from region to region, it is of interest to evaluate the accuracy of satellite-derived rainfall products at the basin scale using ground-based observations. In this study, we evaluated and demonstrated the performance of the three-run GPM-IMERG (early, late, and final) and CHIRPS rainfall datasets against the ground-based observations over the Ziway Lake Basin in Ethiopia. We performed the analysis at monthly and seasonal time scales from 2000 to 2014, using multiple statistical evaluation criteria and graphical methods. While both GPM-IMERG and CHIRPS showed good agreement with ground-observed rainfall data at monthly and seasonal time scales, the CHIRPS products slightly outperformed the GPM-IMERG products. The study thus concluded that CHIRPS or GPM-IMERG rainfall data can be used as a surrogate in the absence of ground-based observed rainfall data for monthly or seasonal agro-hydrological studies.

Keywords: CHIRPS; GPM-IMERG; rainfall data scarcity; agro-hydrology; Rift Valley Lake Basin

1. Introduction

Climate change and variability trend analyses need consistent and long-term time series climate data [1–8] that are required to study the impact of climate change on the agro-hydrological system [9–11]. Such climate studies can benefit from the freely available Global Climate Models (GCMs) outputs such as rainfall data. In addition, complete and long-term rainfall data with high spatial and temporal resolutions are of importance for water resources planning and optimization of crop water productivity especially in water-scarce areas [12–19].

The application of the GCMs rainfall data requires long-term observed-rainfall data for the downscaling and bias correction of coarse resolutions GCMs products into fine resolutions [9,10]. Ground-based rainfall measurement is the most common approach and well recognized as an accurate dataset [20,21]. However, records from the ground-based station are inconsistent over several parts of the world, including Ethiopia [22,23].

Furthermore, available weather stations are inadequate and unevenly distributed to capture rainfall spatial heterogeneity, including less accessibility in remote areas [1,24]. This is a prominent problem, especially in developing countries, including the Ziway Lake Basin [25,26].

The advancement and application of remote sensing technologies offer the possibility of using remotely sensed rainfall data in places where ground-based observed rainfall data are not available [24,27–31]. Several satellite-based rainfall products have been developed with promising approaches for obtaining rainfall estimates at regional and global scales, including blending the ground-based observed rainfall data with remotely sensed data [32]. Some of those satellite-based rainfall products include Tropical Precipitation Measuring Mission Multi-Satellite Precipitation Analysis (TMPA) [33], Precipitation Estimation from Remote Sensed Information using Artificial Neural Networks (PERSIAN) [34], Climate Hazards Infrared Precipitation with Stations (CHIRPS) [35], and Global Precipitation Measurement Integrated Multi-SatellitE Retrieval (GPM-IMERG) [36,37].

Globally, several researchers have evaluated the performance of GPM-IMERG rainfall data using ground-based observations or other existing satellite-based rainfall products [28,38–41]. For example, Tong et al. [38] evaluated the monthly performance of the GPM-IMERG rainfall product using gauge observations at both grid and basin scales for the Nanliu River Basin, Beibu Gulf (Southern coast of China). They concluded that the IMERG showed a high accuracy when detecting light rainfall. Anjum et al. [28] demonstrated IMERG-final run rainfall product estimates by comparing it with gauges and TMPA-based real-time data over the northern highlands of Pakistan at annual, monthly, seasonal, and daily time scale. Their study report showed that the IMERG-final run reasonably well performed than the TMPA-based rainfall estimates. Morsy et al. [40] compared TMPA and IMERG rainfall datasets in the arid environment of El-Qaa Plain, Sinai. They concluded that the IMERG data exhibit superior performance than TMPA in all rainfall intensities. Similarly, Kawo et al. [41] evaluated GPM-IMERG early and late run rainfall estimates with ground gauged rainfall at monthly and seasonal time scales over the Lake Hawassa catchment, Ethiopia. They found that both IMERG-early and late run captured the observed rainfall patterns and values during the rainy season than the dry season.

Many studies have also evaluated the performance of CHIRPS and compared it with ground-based observations at different spatial and temporal scales [31,42–50]. For instance, Wu et al. [50] evaluated the performance of the CHIRPS rainfall dataset against ground-based observed rainfall data over the Yunnan Province, China at monthly, annual, and seasonal scales. They found that CHIRPS data performed well in estimating annual and monthly precipitation. Luo et al. [43] evaluated TRMM and CHIRPS rainfall products in the Lower Lancang-Mekong River Basin. They reported that TRMM rainfall products outperformed the CHIRPS rainfall products. Further, Taye et al. [44] evaluated the performance of CHIRPS and Multi-Source Weighted-Ensemble Precipitation (MSWEP) at a monthly time scale over the upper Blue Nile Basin, Ethiopia. They found that CHIRPS better simulated the magnitude of drought than MSWEP in the different elevation zones of the Upper Blue Nile Basin. Goshime et al. [46] conducted a performance evaluation of CHIRPS rainfall product with the gauged rainfall at monthly and daily temporal resolutions over the Lake Ziway Basin, Ethiopia, and concluded that CHIRPS performed better at the monthly time scale. While several studies have been conducted on evaluating the performance of IMERG and CHRIPS, the previous studies have not simultaneously evaluated and compared the performance of the three IMERG runs (early, late, and final) and CHIRPS at different time scales (monthly and seasonal). Therefore, evaluating and comparing the performance of the recently available different rainfall products at two-time scales is of interest for in-depth and better understanding of their performance and appropriately choosing them as a surrogate when ground-based rainfall observations are lacking. Such studies might also help to identify at what time resolution the satellite-based rainfall estimates can appropriately be used as they play a key role in simulating long-term agro-hydrological modeling and in forecasting changes in freshwater supply and agricultural crop yields [51,52]. Thus, the objectives of this study were to evaluate the accuracy of the satellite-based areal rainfall data over the Ziway Lake Basin at different time scales. We evaluated and compared the CHIRPS and GPM-IMERG of early, late, and final runs with the ground-based observed rainfall data from 12 gauging stations. The evaluation was performed at monthly and seasonal time scales from 2000 to 2014. This study might be useful for the alternative application of remotely sensed precipitation products in simulating the agro-hydrological modeling and climate change trend assessment of the Ziway Lake Basin and elsewhere with similar agro-hydrological conditions, in the Central Rift Valley Lake Basin of Ethiopia.

2. Data and Methods

2.1. Study Area Description

Lake Ziway Basin (LZB) is located between $38°00'-39°30'$ East longitude and $7°00'-8°30'$ North latitude in the Adami Tullu-Jiddo Kombolcha Woreda of the East Shewa Zone, Oromia region, Ethiopia. The basin is about 150 km south of the capital city, Addis Ababa. The town of Ziway (recently named Batu) is situated on the lake's western shore. The altitude of Lake Ziway is approximately 1636 m above mean sea level (amsl), with a maximum water depth of 4 m, a total basin area of about 7300 km^2 (Figure 1) and a lake volume of 1.5 million cubic meters [53]. The majority of the basin is characterized by low to moderately undulating topography but bounded by a steep slope and abrupt faults in the eastern and southeastern escarpments, ranging from 4200 to 1600 m (Figure 1). Lake Ziway Basin experiences the monsoon agro-climate zone characteristics. The rainfall patterns are generally affected by the annual oscillation of the inter-tropical convergence zone that forms wet summer from June to September [54]. The mean annual rainfall of the basin spatially varies from 500 to 1150 mm, with a noticeable temporal variation at a monthly time scale. The mean annual temperature ranges from approximately 15 °C for the highlands to 25 °C close to the lake.

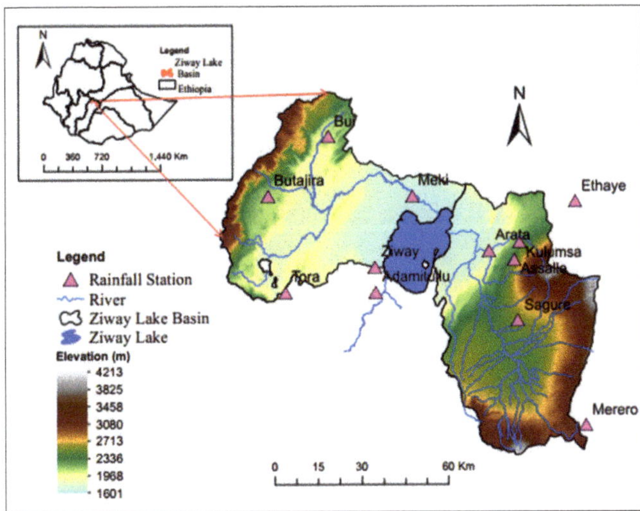

Figure 1. A map of the Ziway Lake Basin, including elevation, rivers, rainfall stations, and Lake Ziway itself.

2.2. Data

2.2.1. Ground Observed Data

In this study, the monthly and seasonal rainfall ground-based observed data from 2000 to 2014 were used as a point of reference for evaluating the CHIRPS and GPM-IMERG. We obtained the data from the Ethiopian National Meteorological Agency (NMA). We originally obtained nineteen climate stations distributed over the Ziway Lake basin with different elevation. However, after performing quality and checking consistency of the data, we selected 12 stations that had good quality and consistent temporal coverage (Table 1). Then, we applied Thiessen polygon method in order to calculate the areal weighted rainfall values of the Ziway Lake Basin (ZLB) from the 12 selected stations. Such approach accounts for the areal coverage of each rain gauge station, the spatial distribution and variability of rainfall for the basin [55]. The areal coverage (Thiessen polygon) of the 12 stations is shown in Figure 2.

Table 1. List of the twelve rainfall stations over the Ziway Lake Basin.

Station Name	Latitude (in Degree)	Longitude (in Degree)	Elevation (m)
Adamitulu	7.86	38.70	1653
Arata	7.98	39.06	1777
Assela	7.96	39.14	2413
Bui	8.33	38.55	2020
Butajira	8.15	38.37	2000
Etheya	8.13	39.33	2129
Kulumsa	8.01	39.16	2211
Meki	8.15	38.82	1662
Merero	7.45	39.37	2940
Sagure	7.77	39.15	2480
Tora	7.86	38.42	2001
Ziway	7.93	38.70	1640

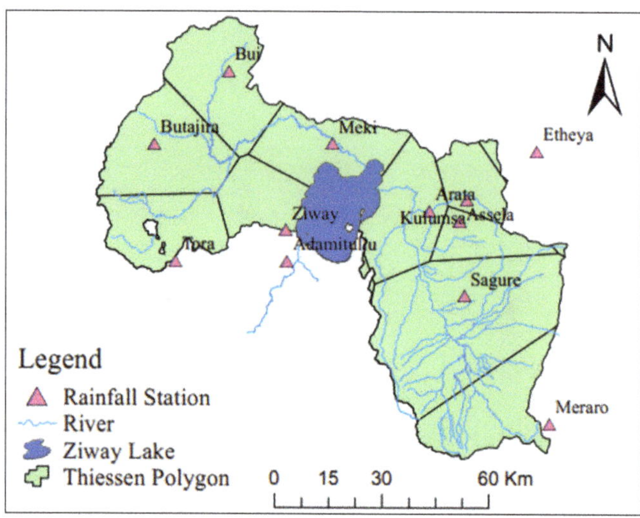

Figure 2. Thiessen Polygon network of the Ziway Lake Basin.

2.2.2. Satellite Precipitation Products

In this study, we considered and evaluated two Satellite Precipitation Products (SPPs). These are CHIRPS and GPM-IMERG.

CHIRPS Database

CHIRPS was launched in early 2014 by the Climate Hazards Group at the University of California, Santa Barbara (UCSB). The CHIRPS precipitation dataset globally covers 50° S–50° N with a horizontal resolution of 0.05° for both daily and monthly time scales. CHIRPS datasets were originally developed to support the United States Agency for

International Development Famine Early Warning Systems Network (FEWS NET) [35] and African Rainfall Climatology [55,56]. Nowadays, the CHIRPS dataset is available in two sets of spatial resolutions i.e., $0.25° \times 0.25°$ and $0.05° \times 0.05°$ from 1981 to the present.

The CHIRPS dataset is developed based on a blend of three data sources [35]: (i) the Climate Hazards Precipitation Climatology (CHPclim) [57], a global precipitation climatology at $0.05°$ latitude and longitude resolution (estimated for each month based on station data, averaged satellite observations, elevation, latitude and longitude) [35,58]; (ii) quasi-global geostationary Thermal Infrared Radiation (TIR) satellite observations, TMPA 3B42 product [33], and (iii) atmospheric model precipitation fields from the National Oceanic and Atmospheric Administration (NOAA) Climate Forecast System (CFS) version 2.0 [59].

According Funk et al. [35], the CHIRPS algorithm encompasses four development processes: (i) a pentad (5 day) rainfall estimate, which is generated from the three-hourly quasi-global geostationary TIR data of Climate Prediction Center (CPC) and the National Climatic Data Center; (ii) a TMPA-3B42 rainfall product, which is used to calibrate the IR pentad estimate; (iii) the calibrated IR pentad product is then multiplied with the Climate Hazards Precipitation Climatology and subsequently divided by the long-term mean to produce the Climate Hazards Group (CHG) IR Precipitation (CHIRP) data; (iv) the pentadal CHIRP values are disaggregated to daily precipitation estimates based on the daily NOAA Climate Forecast System (CFS) fields rescaled to $0.05°$ resolution. Finally, CHIRPS is produced through blending the rainfall stations with the CHIRP data sets and using a modified inverse distance-weighted algorithm [35].

The CHIRPS datasets include rainfall information from a large number of gauges, which is about 1200 stations globally. It should be mentioned that a relatively large number of rain gauge stations were used in East Africa [35]. More than 50 rain gauge stations from the Ethiopian NMA were blended with the CHIRPS products for up-to-date evaluations of the rainfall conditions throughout the major growing seasons of the country. The 50 stations are updated every 10 days [60] and used to correct the CHIRPS datasets [35,49,61] Detailed information regarding the CHIRPS rainfall products was provided in Funk et al. [35]. In this study, we used a higher resolution CHIRPS dataset with a spatial resolution of $0.05° \times 0.05°$ and a daily time scale, which was freely downloaded from (ftp://ftp.chg.ucsb.edu/pub/org/chg/products/CHIRPS-2.0/).

IMERG Database

The GPM-IMERG algorithm combines information from the GPM satellite group to estimate precipitation over the majority of the Earth's surface. The GPM-IMERG was launched by the National Aeronautics and Space Administration (NASA) and the Japan Aeronautics and Exploration Agency (JAXA) in 2014 [62]. This algorithm is particularly valuable over the majority of the Earth's surface that lacks precipitation-measuring instruments on the ground. In the latest release of IMERG (Version 06; V06), the algorithm fuses the early precipitation estimates based on the TRMM satellite (2000−2014) with more recent precipitation estimates collected during the operation based on the GPM satellite (2014–2021). The three gridded products are commonly used for scientific research and operational purposes. There are three different daily IMERG products, which include IMERG Day 1 Early Run (near real-time with a latency of 6 h), IMERG Day 1 Late Run (reprocessed near real-time with a latency of 18 h), and IMERG Day 1 Final Run (gauged-adjusted with a latency of four months) products. In this study, we used the three IMERG products (IMERG-early IMERG-late and IMERG-final run products, with a fine spatial resolution $(0.1° \times 0.1°)$, a high temporal resolution (30 min), and a spatial coverage from 60° S to 60° N, which was freely downloaded from (https://giovanni.gsfc.nasa.gov/giovanni/ (accessed on 4 February 2021)).

2.3. Performance Evaluation Criteria

To identify the best datasets in the study area, we evaluated the performance of CHIRPS and three IMERG (early, late, and final) products against the ground-based rainfall data. We evaluated the monthly and seasonal time scale. We obtained monthly and seasonal rainfall by adding up the daily values on a monthly and seasonal basis in Microsoft Excel 2019 [63], Jupyter Notebook and ArcMap used to visualize data. In Ethiopia, the climate varies mostly with altitude. The lowland areas have hot and arid climatic conditions while plateau areas experience a cold climate, and the season category does not constant over the regions [64,65]. Therefore, in this study, we characterized the performance of CHIRPS and IMERG rainfall datasets for the four seasons of the ZLB. These include *Kiremt* (summer; from June to August), *Tseday* (spring; from September to November), *Bega* (winter; from December to February), and *Belg* (Autumn; from March to May). Then, we evaluated the temporal variations of rainfall for each product.

We consistently used four statistical metrics that include Percent Bias (PBIAS), Root Mean Square Error (RMSE), Nash–Sutcliffe Efficiency (NSE), and Pearson linear Correlation Coefficient (r) to quantitatively compare the performance of the CHIRPS and the three GPM-IMERG rainfall products. PBIAS describes the systematic bias of the CHIRPS and IMERG products. Positive values of PBIAS indicate an overestimation of the rainfall quantity, whereas negative values show an underestimation of the rainfall quantity [28,66,67]. RMSE measures the absolute error magnitude of the CHIRPS and IMERG products, with the smaller the RMSE value, the closer the CHIRPS and IMERG measurements to the ground-observed rainfall. NSE is a normalized statistic that determines the relative magnitude of the residual variance compared to the measured data variance. NSE values range between $-\infty$ and 1, with value 1 indicating a perfect fit between the satellite-based and observed rainfall [42,68]. The degree of linear correlation between the CHIRPS and IMERG and the ground-based rainfall evaluated with r values ranging from -1 to 1 r value of 0 indicates no correlation between the CHIRPS and IMERG products and the observed rainfall. On the other hand, r values of 1 and -1 show perfect positive and negative correlations, respectively [69,70], as summarized in (Table 2). In addition to statistical metrics, we used graph for comparison of SPPs and observed rainfall.

Table 2. List of the statistical metrics, used for the evaluation of satellite rainfall products.

Evaluation Metrics	Description	Equation	Unit	Range	Best Value
Percent Bias (PBIAS)	Measure the average tendency of the SPPs	$PBIAS = \sum_{1=1}^{n} (P_{Si} - P_{Gi}) \frac{1}{\sum_{i=1}^{n} P_S} \times 100$	NA	$(\infty{\sim}\infty)$	0
Root Mean Square (RMSE)	Measure the average magnitude of errors	$RMSE = \sqrt{\sum_{i=1}^{n} (P_{Si} - P_{Gi})^2 \times \frac{1}{N}}$	mm	$[0{\sim}\infty)$	0
Nash–Sutcliffe Efficiency (NSE)	Determines the magnitude of the residual variance	$NSE = 1 - \frac{\sum_{i=1}^{n} (P_{Si} - P_{Gi})^2}{\sum_{i=1}^{n} (P_{Gi} - P_{Gmean})^2}$	NA	$(\infty{\sim}1]$	1
Correlation Coefficient (r)	Indicate the relationship between observed rainfall data and the SPPs products	$r = \frac{\sum_{i=1}^{n} (P_{Gi} - P_{Gmean}) \sum_{i=1}^{n} (P_{Si} - P_{Smean})}{\sqrt{(P_G - P_{Gmean})^2} \sqrt{(P_S - P_{Smean})^2}}$	NA	$[-1{\sim}1]$	1

where: P_{Si} is rainfall from satellite and P_{Gi} the observed rainfall at ith time step (daily, weekly, monthly, or seasonal) with N pairs of data, P_{Gmean} and P_{Smean} are mean observed rainfall and mean satellite rainfall, respectively.

3. Results and Discussion

3.1. Spatial Rainfall Pattern Evaluation

The Ziway Lake Basin seasonal average rainfall distribution of the CHIRPS and IMERG map was compared visually from the 2000–2014 period. Figure 3 shows the seasonal average rainfall distribution for the main rainy (summer) and dry (winter) seasons.

In summer (Figure 3a,b), both CHIRPS and IMERG show that the western part of the basin, which is the eastern highlands of Gurage Zone, receives more rainfall than the eastern part of the basin, which is the western highlands of the Arsi Zone. The spatial rainfall distribution of both CHRIPS and IMERG is consistent with ground-observed rainfall [64]. During the winter season (DJF), a similar rainfall pattern was observed in the western and eastern parts of the basin (Figure 3c,d). Up to 105 mm of rainfall amount is received for the eastern and western part of the basin whereas the central and southern part of the basin receives rainfall up to 45 mm. Overall, both CHRIPS and IMERG showed a decreasing rainfall pattern towards the center i.e., from west to the central part of Ziway Lake Basin (lowland). According to Hailesilassie et al. [64], the observed rainfall is mainly concentrated in the southern and western parts of the basin, while the eastern and central rift valley (low land areas) where the lake is located generally experience low rainfall amounts. CHIRPS relatively well captured that pattern when compared to IMERG, which is probably due to its high spatial resolution and blending of more stations' data [47].

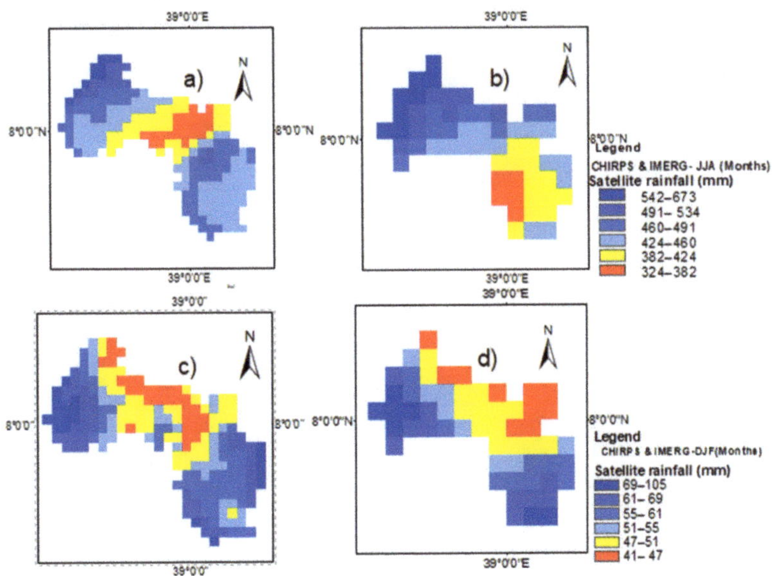

Figure 3. Spatial distribution of main rainy and dry season rainfall (**a,c**) for CHIRPS, (**b,d**) for IMERG for the period 2000–2014.

3.2. Monthly Rainfall Evaluation

Comparison of the CHIRPS and IMERG (early, late, and final run) monthly rainfall data showed a good performance over the Ziway Lake Basin. CHIRPS rainfall generally showed a stronger correlation with the observed rainfall when compared to the three-run IMERG's rainfall (Table 3). The Correlation Coefficient between the early, late, and final IMERG run rainfall and the observed rainfall was high i.e., 0.93, 0.92, and 0.85, respectively. Compared with all IMERG (early, late, and final) products, CHIRPS products showed the highest Correlation Coefficient (0.96) and low Percent Bias (2.22%). In comparison with the IMERG products, the monthly CHIRPS product relatively better represented the ground-observed rainfall values over ZLB with relatively higher r and NSE; and lower RMSE and RBIAS. This is consistent with the previous studies of that confirmed the applicability of CHIRPS precipitation datasets at a monthly time scale in ground-observed data-scarce regions [22,31,46,49,64].

Table 3. Monthly statistical performance evaluation satellite rainfall products for the Ziway Lake Basin.

SPPs	r	NSE	RMSE (mm)	PBIAS (%)
CHIRPS	0.96	0.92	17.45	2.22
IMERG-E	0.92	0.72	28.19	9.67
IMER-L	0.93	0.76	26.12	8.48
IMERG-F	0.85	0.60	34.47	13.0

Figure 4a shows the monthly rainfall values while Figure 4b, c shows the cumulative and scatter values, respectively. The CHIRPS and IMERG-L rainfall product showed the best performance to capture the temporal pattern of monthly rainfall. However, both IMERG-E and IMERG-F products did not well capture the temporal variability of observed rainfall over the study area, indicating that both somehow overestimated the observed rainfall values. As visualized from the cumulative rainfall (Figure 4b), the CHIRPS and IMERG-L captured the monthly cumulative observed rainfall values. The IMERG-E and IMERG-F run smoothly captures the temporal cumulative observed rainfall compared to the CHIRPS and IMERG-L product. As the scatter plot (Figure 4c) indicated, the monthly CHIRPS and IMERG-L rainfall values are close to the monthly observed rainfall values. The CHIRPS data showed capability to represent the monthly maximum observed values compared to all the IMERG's runs. IMERG-L data generally outperformed the IMERG-E and IMERG-F data (Table 3).

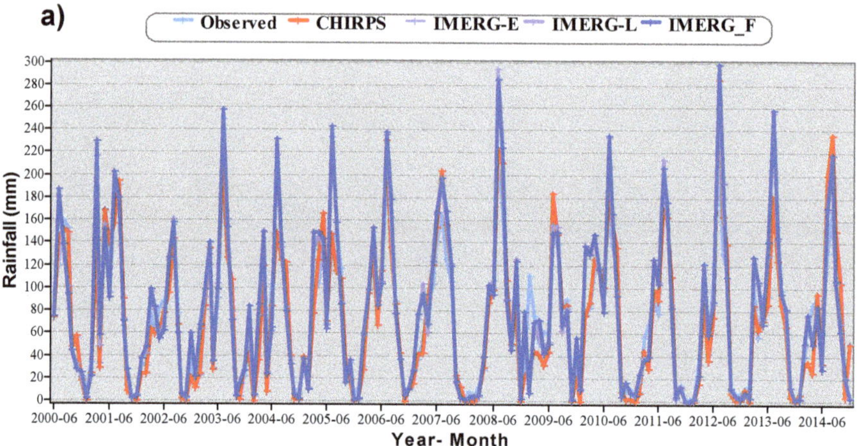

Figure 4. *Cont.*

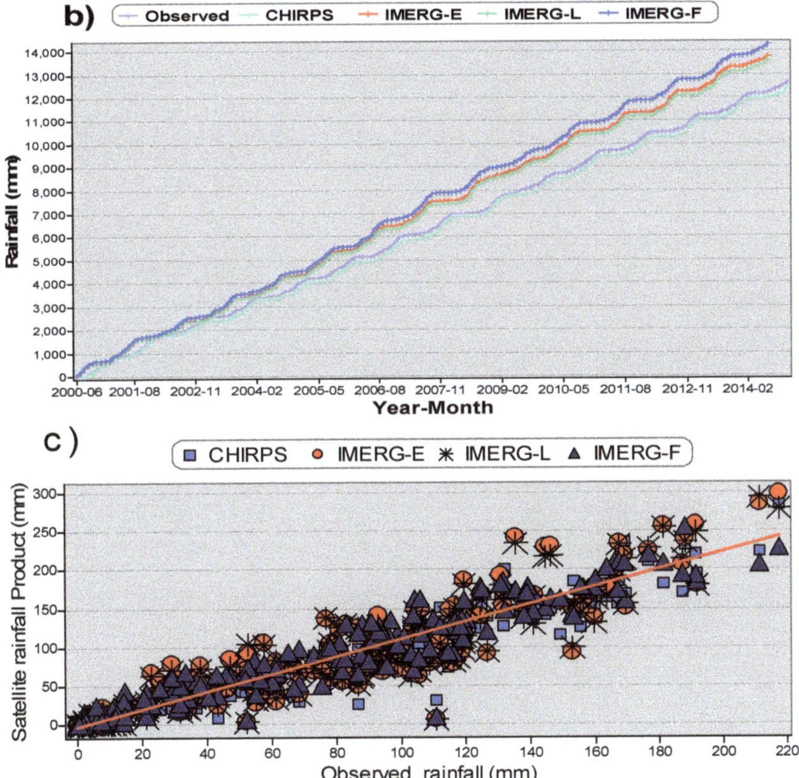

Figure 4. Monthly areal rainfall (**a**), cumulative rainfall depths (**b**), the correlation between monthly satellite-derived (CHIRPS and IMERG-(early, late, and final) run) and observed rainfall (**c**) from (2000–2014) over the Ziway Lake Basin.

3.3. Seasonal Rainfall Evaluation

Figure 5 shows statistical metrics used for seasonal rainfall evaluation of the SPPs versus the ground stations. There were some slight differences between these products on r, RMSE, NSE, and PBIAS (Figure 5a–d). The figure shows that the CHIRPS, the IMERG-E, IMERG-L, and IMERG-F performed well. Moreover, the IMERG-E, IMERG-L, and IMERG-F performance indicated a better relationship during the summer season with an r and NSE values of (0.96 and 0.9 and (0.95 and 0.96), respectively, whereas CHIRPS well-performed with a high r value of 0.92 and low bias error (−2.6) (Figure 5a–d). The three IMERG runs underestimated the summer rainfall by −2.9% to −10%, while CHIRPS underestimated the summer season rainfall by −12% (Figure 5d). All IMERG runs overestimated observed rainfall by 4% to 9.7% in the winter season, whereas CHIRPS underestimated the observed values by −2.6% (Figure 5d). When compared to IMERG runs, CHIRPS achieved higher correlations with observed rainfall during spring, winter, and autumn seasons with r values of 0.93, 0.97, and 0.93 (Figure 5a), respectively. The RMSE values indicated that the CHIRPS data relatively had a small value compared to all IMERG runs, especially during the winter and autumn seasons (Figure 5b). During the spring season, the three IMERG runs had the same r values (0.92) and CHIRPS had (0.93) (Figure 5a).

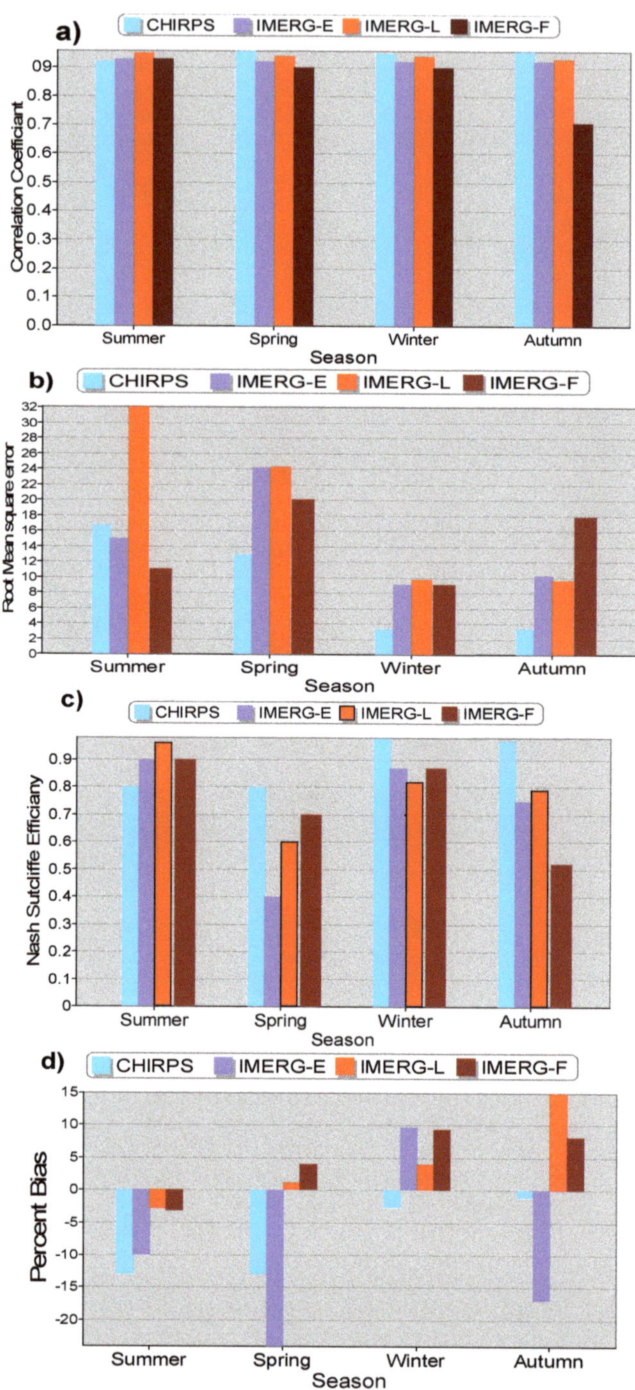

Figure 5. Seasonal performance evaluation indices of CHIRPS, IMERG-E, IMERG-L, and IMERG-F run: Correlation Coefficient (**a**), Root Mean Square Error (**b**), Nash–Sutcliffe Efficiency (**c**), and Percent Bias (**d**) for the period 2000–2014.

A number of previous studies reported the good performance of SPPs at monthly time scales [25,28,41,46,50,68–70]. In general, CHIRPS showed slightly better performance than the other three IMERG runs for monthly and seasonal time scales. Previous studies have already confirmed the superiority of CHIRPS than IMERG runs for different parts of the world [40,71–73], including Ethiopia [31,47,49]. For example, Wedajo et al. [47] reported better rainfall estimation by CHIRPS compared with IMERG and TAMSAT3 and 3B42/3 products for the Dhidhessa River Basin, Ethiopia. Dinku et al. [49] reported better rainfall estimation capability of CHIRPS for east Africa compared to the African Rainfall Climatology version 2 (ARC2) and TAMSAT3 products. The better performance of CHIRPS has been attributed to the capability of the algorithm to integrate satellite, rain gauges, and reanalysis products, combined with its higher spatial and temporal resolutions than IMERG products [35].

Overall, the statistical evaluation results indicate that both CHIRPS and IMERG are capable of estimating and detecting observed monthly and seasonal rainfall values of the ZLB. Therefore, the monthly and seasonal CHIRPS and IMERG-F data are a reliable source for simulating monthly and seasonal agro-hydrological processes, estimating the seasonal crop water requirement, and accounting the stocks and fluxes of water in the Ziway Lake Basin.

4. Conclusions

In this study, we evaluated and compared the performance of IMERG and CHIRPS rainfall products against ground-observed rainfall data over the Ziway Lake Basin. The analyses covered the period from 2000 to 2014 at monthly and seasonal time scales. We used four statistical evaluation parameters: Correlation Coefficient, Nash–Sutcliffe Efficiency, Percent Bias, and Root Mean Square Error. The two rainfall products performed well for both monthly and seasonal time scales. Overall, while the CHIRPS's rainfall datasets showed slightly better performance over the IMERG's datasets, both datasets can be used at a monthly or coarser temporal resolution when ground-based rainfall data are not available. This can greatly contribute to continuous spatiotemporal monitoring of drought and helping the water managers and agricultural planners implementing mitigation measures and improving the livelihood of the stakeholders in the basin.

The follow up research should focus on the evaluation and comparison of the grid point satellite dataset with interposed ground station data, considering point to point performance evaluation at daily time basis. Future evaluation studies should also include the Climate Hazards Group Infrared Precipitation (CHIRP) satellite-only product.

Author Contributions: Conceptualization, A.T.H.; data curation and analysis, A.T.H. and N.S.K.; Formula analysis and Methodology, A.T.H.; writing–original draft, A.T.H.; writing–review, A.T.H.; editing, A.T.H., O.T.L. and A.D.C.; supervision O.T.L., T.A. and A.D.C. All authors have read and agreed to published version of the manuscript.

Funding: This work was financially supported by the Africa Center of Excellence for Water Management, Addis Ababa University.

Acknowledgments: The authors would like to thanks the National Meteorological Agency (NMA) of Ethiopia for providing the weather data. We would also like to express our sincere gratitude to the Africa Centre of Excellence for Water Management, Addis Ababa University for the support to conduct this research.

Conflicts of Interest: The authors declare no conflict of interest.

References

1. Nikolopoulos, E.I.; Destro, E.; Maggioni, V.; Marra, F.; Borga, M. Satellite Rainfall Estimates for Debris Flow Prediction: An Evaluation Based on Rainfall Accumulation–Duration Thresholds. *J. Hydrometeorol.* **2017**, *18*, 2207–2214. [CrossRef]
2. Intergovernmental Panel on Climate Change (IPCC). Summary for Policymakers. Climate change 2014: Part of the Working Group III Contribution to the Fifth Assessment Report of the IPCC, Geneva, Switzerland. Available online: http://pure.iiasa.ac.at/11125 (accessed on 20 April 2021).
3. Hoegh-Guldberg, O.; Jacob, D.; Bindi, M.; Brown, S.; Camilloni, I.; Diedhiou, A.; Djalante, R.; Ebi, K.; Engelbrecht, F.; Guiot, J.; et al. Impacts of 1.5 C Global Warming on Natural and Human Systems. *Global warming of 1.5 C. An IPCC Special Report.* Available online: http://hdl.handle.net/10138/311749 (accessed on 15 June 2021).
4. Bayissa, Y.; Tadesse, T.; Demisse, G.; Shiferaw, A. Evaluation of Satellite-Based Rainfall Estimates and Application to Monitor Meteorological Drought for the Upper Blue Nile Basin, Ethiopia. *Remote. Sens.* **2017**, *9*, 669. [CrossRef]
5. Asfaw, A.; Simane, B.; Hassen, A.; Bantider, A. Variability and time series trend analysis of rainfall and temperature in northcentral Ethiopia: A case study in Woleka sub-basin. *Weather. Clim. Extrem.* **2018**, *19*, 29–41. [CrossRef]
6. Ayalew, D. Variability of rainfall and its current trend in Amhara region, Ethiopia. *Afr. J. Agric. Res.* **2012**, *7*, 1475–1486. [CrossRef]
7. Pachauri, R.K.; Allen, M.R.; Barros, V.R.; Broome, J.; Cramer, W.; Christ, R.; Church, J.A.; Clarke, L.; Dahe, Q.; Dasgupta, P.; et al. Climate change 2014: Synthesis report. Contribution of Working Groups I, II and III to the fifth assessment report of the Intergovernmental Panel on Climate Change. *Ipcc* **2014**. [CrossRef]
8. Chartzoulakis, K.; Bertaki, M. Sustainable Water Management in Agriculture under Climate Change. *Agric. Agric. Sci. Procedia* **2015**, *4*, 88–98. [CrossRef]
9. Emami, F.; Koch, M. Agricultural Water Productivity-Based Hydro-Economic Modeling for Optimal Crop Pattern and Water Resources Planning in the Zarrine River Basin, Iran, in the Wake of Climate Change. *Sustainability* **2018**, *10*, 3953. [CrossRef]
10. Hassan, I.; Kalin, R.M.; White, C.J.; Aladejana, J.A. Selection of CMIP5 GCM Ensemble for the Projection of Spatio-Temporal Changes in Precipitation and Temperature over the Niger Delta, Nigeria. *Water* **2020**, *12*, 385. [CrossRef]
11. Rakhimova, M.; Liu, T.; Bissenbayeva, S.; Mukanov, Y.; Gafforov, K.S.; Bekpergenova, Z.; Gulakhmadov, A. Assessment of the Impacts of Climate Change and Human Activities on Runoff Using Climate Elasticity Method and General Circulation Model (GCM) in the Buqtyrma River Basin, Kazakhstan. *Sustainability* **2020**, *12*, 4968. [CrossRef]
12. Huntingford, C.; Lambert, F.H.; Gash, J.; Taylor, C.M.; Challinor, A.J. Aspects of climate change prediction relevant to crop productivity. *Philos. Trans. R. Soc. B Biol. Sci.* **2005**, *360*, 1999–2009. [CrossRef]
13. Fischer, G.; Tubiello, F.N.; van Velthuizen, H.; Wiberg, D. Climate change impacts on irrigation water requirements: Effects of mitigation, 1990–2080. *Technol. Forecast. Soc. Chang.* **2007**, *74*, 1083–1107. [CrossRef]
14. Ficklin, D.; Luo, Y.; Luedeling, E.; Zhang, M. Climate change sensitivity assessment of a highly agricultural watershed using SWAT. *J. Hydrol.* **2009**, *374*, 16–29. [CrossRef]
15. Legesse, S.; Tadele, K.; Mariam, B.G. Potential Impacts of Climate Change on the Hydrology and Water resources Availability of Didessa Catchment, Blue Nile River Basin, Ethiopia. *J. Geol. Geosci.* **2015**, *4*. [CrossRef]
16. Jia, K.; Ruan, Y.; Yang, Y.; Zhang, C. Assessing the Performance of CMIP5 Global Climate Models for Simulating Future Precipitation Change in the Tibetan Plateau. *Water* **2019**, *11*, 1771. [CrossRef]
17. Gedefaw, M.; Wang, H.; Yan, D.; Qin, T.; Wang, K.; Girma, A.; Batsuren, D.; Abiyu, A. Water Resources Allocation Systems under Irrigation Expansion and Climate Change Scenario in Awash River Basin of Ethiopia. *Water* **2019**, *11*, 1966. [CrossRef]
18. Kang, Y.; Khan, S.; Ma, X. Climate change impacts on crop yield, crop water productivity and food security—A review. *Prog. Nat. Sci.* **2009**, *19*, 1665–1674. [CrossRef]
19. Worku, T.; Khare, D.; Tripathi, S.K. Spatiotemporal trend analysis of rainfall and temperature, and its implications for crop production. *J. Water Clim. Chang.* **2019**, *10*, 799–817. [CrossRef]
20. Dawit, M.; Halefom, A.; Teshome, A.; Sisay, E.; Shewayirga, B.; Dananto, M. Changes and variability of precipitation and temperature in the Guna Tana watershed, Upper Blue Nile Basin, Ethiopia. *Model. Earth Syst. Environ.* **2019**, *5*, 1395–1404. [CrossRef]
21. Mohammed, Y.; Yimer, F.; Tadesse, M.; Tesfaye, K. Variability and trends of rainfall extreme events in north east highlands of Ethiopia. *Int. J. Hydrol.* **2018**, *2*. [CrossRef]
22. Gebere, S.B.; Alamirew, T.; Merkel, B.J.; Melesse, A.M. Performance of High Resolution Satellite Rainfall Products over Data Scarce Parts of Eastern Ethiopia. *Remote. Sens.* **2015**, *7*, 11639–11663. [CrossRef]
23. Salerno, J.; Diem, J.E.; Konecky, B.L.; Hartter, J. Recent intensification of the seasonal rainfall cycle in equatorial Africa revealed by farmer perceptions, satellite-based estimates, and ground-based station measurements. *Clim. Chang.* **2019**, *153*, 123–139. [CrossRef]
24. Karger, D.N.; Schmatz, D.R.; Dettling, G.; Zimmermann, N.E. High-resolution monthly precipitation and temperature time series from 2006 to 2100. *Sci. Data* **2020**, *7*, 1–10. [CrossRef]
25. Al-Wagdany, A.S. Inconsistency in rainfall characteristics estimated from records of different rain gauges. *Arab. J. Geosci.* **2016**, *9*, 1–10. [CrossRef]
26. Musie, M.; Sen, S.; Chaubey, I. Hydrologic Responses to Climate Variability and Human Activities in Lake Ziway Basin, Ethiopia. *Water* **2020**, *12*, 164. [CrossRef]

27. Serinaldi, F.; Kilsby, C.G.; Lombardo, F. Untenable nonstationarity: An assessment of the fitness for purpose of trend tests in hydrology. *Adv. Water Resour.* **2018**, *111*, 132–155. [CrossRef]

28. Anjum, M.N.; Ahmad, I.; Ding, Y.; Shangguan, D.; Zaman, M.; Ijaz, M.W.; Sarwar, K.; Han, H.; Yang, M. Assessment of IMERG-V06 Precipitation Product over Different Hydro-Climatic Regimes in the Tianshan Mountains, North-Western China. *Remote. Sens.* **2019**, *11*, 2314. [CrossRef]

29. Xu, M.; Kang, S.; Wu, H.; Yuan, X. Detection of spatio-temporal variability of air temperature and precipitation based on long-term meteorological station observations over Tianshan Mountains, Central Asia. *Atmos. Res.* **2018**, *203*, 141–163. [CrossRef]

30. Tang, G.; Clark, M.P.; Papalexiou, S.M.; Ma, Z.; Hong, Y. Have satellite precipitation products improved over last two decades? A comprehensive comparison of GPM IMERG with nine satellite and reanalysis datasets. *Remote Sens. Environ.* **2020**, *240*, 111697. [CrossRef]

31. Fenta, A.A.; Yasuda, H.; Shimizu, K.; Ibaraki, Y.; Haregeweyn, N.; Kawai, T.; Belay, A.S.; Sultan, D.; Ebabu, K. Evaluation of satellite rainfall estimates over the Lake Tana basin at the source region of the Blue Nile River. *Atmos. Res.* **2018**, *212*, 43–53. [CrossRef]

32. Ma, Z.; Tan, X.; Yang, Y.; Chen, X.; Kan, G.; Ji, X.; Lu, H.; Long, J.; Cui, Y.; Hong, Y. The First Comparisons of IMERG and the Downscaled Results Based on IMERG in Hydrological Utility over the Ganjiang River Basin. *Water* **2018**, *10*, 1392. [CrossRef]

33. Huffman, G.J.; Bolvin, D.T.; Nelkin, E.J.; Wolff, D.B.; Adler, R.F.; Gu, G.; Hong, Y.; Bowman, K.P.; Stocker, E.F. The TRMM Multisatellite Precipitation Analysis (TMPA): Quasi-Global, Multiyear, Combined-Sensor Precipitation Estimates at Fine Scales. *J. Hydrometeorol.* **2007**, *8*, 38–55. [CrossRef]

34. Hsu, K.L.; Gao, X.; Sorooshian, S.; Gupta, H.V. Precipitation estimation from remotely sensed information using artificial neural networks. *J. Appl. Meteorol.* **1997**, *36*, 1176–1190. [CrossRef]

35. Funk, C.; Peterson, P.; Landsfeld, M.; Pedreros, D.; Verdin, J.; Shukla, S.; Husak, G.; Rowland, J.; Harrison, L.; Hoell, A.; et al. The climate hazards infrared precipitation with stations—A new environmental record for monitoring extremes. *Sci. Data* **2015**, *2*, 150066. [CrossRef] [PubMed]

36. Hou, A.Y.; Kakar, R.K.; Neeck, S.; Azarbarzin, A.A.; Kummerow, C.D.; Kojima, M.; Oki, R.; Nakamura, K.; Iguchi, T. The global precipitation measurement mission. *Bull. Am. Meteorol. Soc.* **2014**, *95*, 701–722. [CrossRef]

37. Huffman, G.J.; Bolvin, D.T.; Braithwaite, D.; Hsu, K.; Joyce, R.; Xie, P.; Yoo, S.H. NASA global precipitation meas-urement (GPM) integrated multi-satellite retrievals for GPM (IMERG). *Algorithm Theor. Basis Doc. (ATBD) Version* **2015**, *4*, 26.

38. Tong, K.; Zhao, Y.; Wei, Y.; Hu, B.; Lu, Y. Evaluation and Hydrological Validation of GPM Precipitation Products over the Nanliu River Basin, Beibu Gulf. *Water* **2018**, *10*, 1777. [CrossRef]

39. Xiao, S.; Xia, J.; Zou, L. Evaluation of Multi-Satellite Precipitation Products and Their Ability in Capturing the Characteristics of Extreme Climate Events over the Yangtze River Basin, China. *Water* **2020**, *12*, 1179. [CrossRef]

40. Morsy, M.; Scholten, T.; Michaelides, S.; Borg, E.; Sherief, Y.; Dietrich, P. Comparative Analysis of TMPA and IMERG Precipitation Datasets in the Arid Environment of El-Qaa Plain, Sinai. *Remote. Sens.* **2021**, *13*, 588. [CrossRef]

41. Kawo, N.S.; Hordofa, A.T.; Karuppannan, S. Performance evaluation of GPM-IMERG early and late rainfall estimates over Lake Hawassa catchment, Rift Valley Basin, Ethiopia. *Arab. J. Geosci.* **2021**, *14*, 1–14. [CrossRef]

42. Prakash, S. Performance assessment of CHIRPS, MSWEP, SM2RAIN-CCI, and TMPA precipitation products across India. *J. Hydrol.* **2019**, *571*, 50–59. [CrossRef]

43. Luo, X.; Wu, W.; He, D.; Li, Y.; Ji, X. Hydrological Simulation Using TRMM and CHIRPS Precipitation Estimates in the Lower Lancang-Mekong River Basin. *Chin. Geogr. Sci.* **2019**, *29*, 13–25. [CrossRef]

44. Taye, M.; Sahlu, D.; Zaitchik, B.; Neka, M. Evaluation of Satellite Rainfall Estimates for Meteorological Drought Analysis over the Upper Blue Nile Basin, Ethiopia. *Geoscience* **2020**, *10*, 352. [CrossRef]

45. Pang, J.; Zhang, H.; Xu, Q.; Wang, Y.; Wang, Y.; Zhang, O.; Hao, J. Hydrological evaluation of open-access precipitation data using SWAT at multiple temporal and spatial scales. *Scales Hydrol. Earth Sci.* **2020**, *24*, 3603–3626. [CrossRef]

46. Goshime, D.W.; Absi, R.; Ledésert, B. Evaluation and Bias Correction of CHIRP Rainfall Estimate for Rainfall-Runoff Simulation over Lake Ziway Watershed, Ethiopia. *Hydrology* **2019**, *6*, 68. [CrossRef]

47. Wedajo, G.K.; Muleta, M.K.; Awoke, B.G. Performance evaluation of multiple satellite rainfall products for Dhidhessa River Basin (DRB), Ethiopia. *Atmos. Meas. Tech.* **2021**, *14*, 2299–2316. [CrossRef]

48. Esayas, B.; Simane, B.; Teferi, E.; Ongoma, V.; Tefera, N. Trends in Extreme Climate Events over Three Agroecological Zones of Southern Ethiopia. *Adv. Meteorol.* **2018**, *2018*, 1–17. [CrossRef]

49. Dinku, T.; Funk, C.; Peterson, P.; Maidment, R.; Tadesse, T.; Gadain, H.; Ceccato, P. Validation of the CHIRPS satellite rainfall estimates over eastern Africa. *Q. J. R. Meteorol. Soc.* **2018**, *144*, 292–312. [CrossRef]

50. Wu, W.; Li, Y.; Luo, X.; Zhang, Y.; Ji, X.; Li, X. Performance evaluation of the CHIRPS precipitation dataset and its utility in drought monitoring over Yunnan Province, China. *Geomat. Nat. Hazards Risk* **2019**, *10*, 2145–2162. [CrossRef]

51. Korres, N.; Norsworthy, J.; Burgos, N.; Oosterhuis, D. Temperature and drought impacts on rice production: An agronomic perspective regarding short- and long-term adaptation measures. *Water Resour. Rural. Dev.* **2017**, *9*, 12–27. [CrossRef]

52. Guo, H.; Chen, S.; Bao, A.; Hu, J.; Gebregiorgis, A.S.; Xue, X.; Zhang, X. Inter-comparison of high-resolution satellite precipitation products over Central Asia. *Remote Sens.* **2015**, *7*, 7181–7211. [CrossRef]

53. Legesse, D.; Ayenew, T. Effect of improper water and land resource utilization on the central Main Ethiopian Rift lakes. *Quat. Int.* **2006**, *148*, 8–18. [CrossRef]

54. Alhamshry, A.; Fenta, A.A.; Yasuda, H.; Kimura, R.; Shimizu, K. Seasonal Rainfall Variability in Ethiopia and Its Long-Term Link to Global Sea Surface Temperatures. *Water* **2020**, *12*, 55. [CrossRef]
55. Mair, A.; Fares, A. Comparison of Rainfall Interpolation Methods in a Mountainous Region of a Tropical Island. *J. Hydrol. Eng.* **2011**, *16*, 371–383. [CrossRef]
56. Novella, N.S.; Thiaw, W.M. African Rainfall Climatology Version 2 for Famine Early Warning Systems. *J. Appl. Meteorol. Clim.* **2013**, *52*, 588–606. [CrossRef]
57. Funk, C.; Michaelsen, J.; Marshall, M.T. Mapping Recent Decadal Climate Variations in Precipitation and Temperature across Eastern Africa and the Sahel. *Remote Sens. Drought Innov. Monit. Approaches* **2012**, *24*, 331. [CrossRef]
58. Kisaka, M.O.; Mucheru-Muna, M.; Ngetich, F.K.; Mugwe, J.; Mugendi, D.N.; Mairura, F. Rainfall Variability, Drought Characterization, and Efficacy of Rainfall Data Reconstruction: Case of Eastern Kenya. *Adv. Meteorol.* **2015**, *2015*, 1–16. [CrossRef]
59. Saha, S.; Moorthi, S.; Pan, H.-L.; Wu, X.; Wang, J.; Nadiga, S.; Tripp, P.; Kistler, R.; Woollen, J.; Behringer, D.; et al. The NCEP Climate Forecast System Reanalysis. *Bull. Am. Meteorol. Soc.* **2010**, *91*, 1015–1058. [CrossRef]
60. Nega, W.; Hailu, B.T.; Fetene, A. An assessment of the vegetation cover change impact on rainfall and land surface temperature using remote sensing in a subtropical climate, Ethiopia. *Remote. Sens. Appl. Soc. Environ.* **2019**, *16*, 100266. [CrossRef]
61. Taye, M.T.; Dyer, E.; Hirpa, F.A.; Charles, K. Climate Change Impact on Water Resources in the Awash Basin, Ethiopia. *Water* **2018**, *10*, 1560. [CrossRef]
62. Ma, Z.; Xu, J.; Zhu, S.; Yang, J.; Tang, G.; Yang, Y.; Shi, Z.; Hong, Y. AIMERG: A new Asian precipitation dataset (0.1°/half-hourly, 2000–2015) by calibrating the GPM-era IMERG at a daily scale using APHRODITE. *Earth Syst. Sci. Data* **2020**, *12*, 1525–1544. [CrossRef]
63. Ehsani, M.R.; Behrangi, A.; Adhikari, A.; Song, Y.; Huffman, G.J.; Adler, R.F.; Bolvin, D.T.; Nelkin, E.J. Assessment of the Advanced Very High-Resolution Radiometer (AVHRR) for Snowfall Retrieval in High Latitudes Using CloudSat and Machine Learning. *J. Hydrometeorol.* **2021**, *1*. [CrossRef]
64. Hailesilassie, W.T.; Ayenew, T.; Tekleab, S. Analysing Trends and Spatio-Temporal Variability of Precipitation in the Main Central Rift Valley Lakes Basin, Ethiopia. *Environ. Earth Sci. Res. J.* **2021**, *8*, 37–47. [CrossRef]
65. Conway, D. Some aspects of climate variability in the north east Ethiopian highlands—Wollo and Tigray. *SINET Ethiop. J. Sci.* **2000**, *23*, 139–161. [CrossRef]
66. Gupta, S.K.; Ritchey, N.A.; Wilber, A.C.; Whitlock, C.H.; Gibson, G.G.; Stackhouse, P.W., Jr. A climatology of surface radiation budget derived from satellite data. *J. Clim.* **1999**, *12*, 2691–2710. [CrossRef]
67. Lutz, A.F.; Ter Maat, H.W.; Biemans, H.; Shrestha, A.B.; Wester, P.; Immerzeel, W. Selecting representative climate models for climate change impact studies: An advanced envelope-based selection approach. *Int. J. Clim.* **2016**, *36*, 3988–4005. [CrossRef]
68. McCuen, R.H.; Knight, Z.; Cutter, A.G. Evaluation of the Nash–Sutcliffe Efficiency Index. *J. Hydrol. Eng.* **2006**, *11*, 597–602. [CrossRef]
69. Farlie, D.J.G. The Performance of Some Correlation Coefficients for a General Bivariate Distribution. *Biometrika* **1960**, *47*, 47. [CrossRef]
70. Chicco, D.; Jurman, G. The advantages of the Matthews correlation coefficient (MCC) over F1 score and accuracy in binary classification evaluation. *BMC Genom.* **2020**, *21*, 6. [CrossRef]
71. Rivera, J.A.; Marianetti, G.; Hinrichs, S. Validation of CHIRPS precipitation dataset along the Central Andes of Argentina. *Atmos. Res.* **2018**, *213*, 437–449. [CrossRef]
72. Saeidizand, R.; Sabetghadam, S.; Tarnavsky, E.; Pierleoni, A. Evaluation of CHIRPS rainfall estimates over Iran. *Q. J. R. Meteorol. Soc.* **2018**, *144*, 282–291. [CrossRef]
73. Tramblay, Y.; Thiemig, V.; Dezetter, A.; Hanich, L. Evaluation of satellite-based rainfall products for hydrological modelling in Morocco. *Hydrol. Sci. J.* **2016**, *61*, 2509–2519. [CrossRef]

Article

Modeling the Soil Response to Rainstorms after Wildfire and Prescribed Fire in Mediterranean Forests

Manuel Esteban Lucas-Borja [1], Giuseppe Bombino [2], Bruno Gianmarco Carrà [2], Daniela D'Agostino [2], Pietro Denisi [2], Antonino Labate [2], Pedro Antonio Plaza-Alvarez [1] and Demetrio Antonio Zema [2,*]

[1] Escuela Técnica Superior Ingenieros Agrónomos y Montes, Universidad de Castilla-La Mancha, Campus Universitario, E-02071 Albacete, Spain; ManuelEsteban.Lucas@uclm.es (M.E.L.-B.); pedro.plaza@uclm.es (P.A.P.-A.)

[2] Department AGRARIA, Mediterranean University of Reggio Calabria, Loc. Feo di Vito, I-89122 Reggio Calabria, Italy; giuseppe.bombino@unirc.it (G.B.); brunog.carra@unirc.it (B.G.C.); daniela.dagostino@unirc.it (D.D.); pietro.denisi@unirc.it (P.D.); a.labate@unirc.it (A.L.)

* Correspondence: dzema@unirc.it

Received: 17 November 2020; Accepted: 15 December 2020; Published: 17 December 2020

Abstract: The use of the Soil Conservation Service-curve number (SCS-CN) model for runoff predictions after rainstorms in fire-affected forests in the Mediterranean climate is quite scarce and limited to the watershed scale. To validate the applicability of this model in this environment, this study has evaluated the runoff prediction capacity of the SCS-CN model after storms at the plot scale in two pine forests of Central-Eastern Spain, affected by wildfire (with or without straw mulching) or prescribed fire and in unburned soils. The model performance has been compared to the predictions of linear regression equations between rainfall depth and runoff volume. The runoff volume was simulated with reliability by the linear regression only for the unburned soil (coefficient of Nash and Sutcliffe E = 0.73–0.89). Conversely, the SCS-CN model was more accurate for burned soils (E = 0.81–0.97), also when mulching was applied (E = 0.96). The performance of this model was very satisfactory in predicting the maximum runoff. Very low values of CNs and initial abstraction were required to predict the particular hydrology of the experimental areas. Moreover, the post-fire hydrological "window-of-disturbance" could be reproduced only by increasing the CN for the storms immediately after the wildfire. This study indicates that, in Mediterranean forests subject to the fire risk, the simple linear equations are feasible to predict runoff after low-intensity storms, while the SCS-CN model is advisable when runoff predictions are needed to control the flooding risk.

Keywords: hydrological models; rainfall; surface runoff; linear regression models; curve number; SCS.CN model; mulching; wildfire; prescribed fire

---

1. Introduction

The importance of forests in the climatic context at the planetary scale is well known, since forests produce oxygen and store carbon, regulate water and energy fluxes, support biodiversity and provide other fundamental ecosystem services [1–3]. However, the fundamental role of forests is threatened by some natural and anthropogenic agents, such as the extreme weather events and fire, with a long history of influence on forest ecosystems [4]. Extreme weather events (e.g., heavy storms and severe drought) are more and more intensified by climate change trends and occur in all regions of the world [5,6]. The fire effects extend to several components of forests, such as soil, vegetation, air and surface water [7]. With regard to the effects on surface water and soil, fire strongly alters the hydrological response of recently burnt areas, increasing by many folds the soil's aptitude to generate runoff and erosion compared the unburned forest areas [8,9]. High runoff and erosion rates in forests increase the flooding risk, debris flow occurrence and water quality alteration in downstream areas, with possible loss of human lives and heavy damage to infrastructures and environment [10,11]. For instance, with regard to the Mediterranean climate, exceptionally high erosion rates (up to 100 tons per ha) have been reported immediately after wildfire by Menendez-Duarte et al. [12] in the Iberian peninsula, while Lopez-Batalla et al. [13] measured increases in flood runoff by 30% and in peak discharges by 120% in the same environments. These studies together with a large body of literature (see several examples in the milestone review of Shakesby [14]) clearly demonstrate the need to control and mitigate the hydrological response of forest soils after the wildfire adopting prediction tools and post-fire management actions.

The hydrological processes in forests are influenced by several factors, among which fire severity is important [15]. In other words, the more severe the fire, the more susceptible the soil to increases in runoff and erosion and worsening in water quality changes in the downstream ecosystems [7]. For instance, Lucas-Borja et al. [7] found increases in runoff and erosion in soil burned by wildfire by about 20% and even 200%, respectively, compared to unburned soils in Central-Eastern Spain.

The hydrology of burned forests is very complex, since it is the product of several factors, such as climate and edapho-climatic conditions, fire severity, soil, vegetation, morphology and land management after fire [15–19]. The needs to predict the impacts of fire on runoff and erosion and to implement the most effective actions for rehabilitation of fire-affected areas have increased the demand for hydrological models [16,20]. This demand is particularly important for forest managers working in the Mediterranean Basin, which is characterized by dry and hot summers followed by frequent and high-intensity rains immediately after the wildfire season [15,21]. In Mediterranean areas, increases in wildfire frequency and burned areas are expected under the forecasted climate scenarios [22,23].

In Mediterranean forests, the literature reports applications and verifications of hydrological models with different nature and complexity: from the simplest empirical models (such as the Soil Conservation Service (SCS)-curve number model to predict runoff, the universal soil loss equation, USLE, to simulate soil erosion) through the semiempirical models (e.g., the Morgan–Morgan–Finney model, MMF) until the most complex physically-based models (for instance, the Water Erosion Prediction Project, WEPP) or even the artificial neural networks [24,25]. Nonetheless, the empirical models are sometime more commonly used compared to the more complex models, mainly in data-poor environments (that is, in those situations with limited availability of parameter inputs) and for quick identification of sources of water, sediments and pollutants in forests [24–26].

Accurate predictions of surface runoff are fundamental to achieve reliable estimations of erosion rates and water quality parameters using hydrological models [27], particularly in forests subject to climate change and fire, since the latter factors play a large influence of the soil's hydrological response [28,29]. The Soil Conservation Service (SCS)-curve number (CN) model (hereinafter "SCS-CN model") is one of the most common methods for estimating the runoff volume generated by a rainstorm [30,31]. The popularity of this method is due to its simplicity, ease of use, widespread acceptance and large availability of input data [32]. Moreover, the SCS-CN model takes into account most of the factors that influence runoff generation, such as soil type, land use, hydrologic condition and antecedent moisture of the soil [33]. The model has also been incorporated as a hydrological submodel in several distributed rainfall–runoff models at the watershed scale (e.g., AnnAGNPS—annual agricultural non-point source pollution model, SWAT—soil and water assessment tool model and HEC-HMS—hydrologic engineering center-hydrologic modeling system model), supporting its robustness and popularity [33,34]). To date, there is no any alternative model that offers as many advantages as the SCS-CN model, which therefore is still commonly used in the large majority of environments and climatic conditions [35].

However, various studies conducted throughout the world on the applicability of the SCS-CN method have suggested a need for further improvement or overhauling of the model [32,36], since in some environments the method provides unsatisfactory predictions, particularly when the soil's hydrological response does not follow the runoff generation mechanism by saturation-excess.

Moreover, in spite of the large number of requisites, the model has been surprisingly little used for hydrological predictions in fire-affected areas, and the CN values are not completely known in burned areas [37]. The hydrological research has been mainly carried out at the watershed scale, where post-fire runoff has been predicted using the SCS-CN method incorporated in watershed-scale models (e.g., WILDCAT4 Flow Model [38] and FIRE HYDRO [39]). For instance, the SCS-CN model was used by Candela et al. [40], who analyzed the flood frequency curves for pre- and post-fire conditions, showing an increase in the average curve numbers and a decrease in the catchment time lag. Increases by 25 units in post-fire CNs were estimated by Soulis [37] in a small Greek watershed using pre-fire and post-fire rainfall–runoff datasets. A daily-constant CN in the SWAT model was used by Nunes et al. [41] to simulate the effects of soil water repellency on runoff from burnt hillslopes in a Mediterranean forest throughout three years after fire. It is therefore evident that the modeling experiences are scarce at the plot scale. At this scale, modeling of soil hydrology is less complex compared to the watershed scale, where the hydrological response to Mediterranean storms is further complicated by a combination and overlaying of several hydrological processes (e.g., water routing in the channel network, ponding and uneven soil properties) other than surface runoff generation. Furthermore, the studies about the hydrological effects of post-fire management on runoff in forests using the SCS-CN model are scarce.

Therefore, there is a need of further studies that must evaluate the runoff prediction capacity of the SCS-CN model in forest hillslopes or plots affected by fire of different severity—a fire parameter referring to the effects of wildfire on plant communities—in comparison with simpler models that estimate runoff directly from precipitation, such as the linear regression equations. In other words, is the SCS-CN model accurate to predict surface runoff in Mediterranean burned forests? Is it able to simulate post-fire hydrology with or without rehabilitation management actions (such as straw mulching) after a wildfire? When is its use convenient compared to a simpler linear regression between rainfall and runoff?

This study aims to reply to these questions, evaluating the hydrological performance of the SCS-CN model in two pine forests of Central-Eastern Spain affected by a wildfire and a prescribed fire, respectively, having different fire severity. More specifically, observations of rainfall–runoff patterns collected throughout one year in undisturbed soils (assumed as control) and in plots subject to prescribed fire/wildfire (the latter with or without a mulching treatment) are compared with the corresponding predictions of the SCS-CN model and linear rainfall–runoff regressions. The outcomes of this study help land managers to adopt strategies to control the hydrological effects of fire in Mediterranean forests.

2. Materials and Methods

2.1. Study Areas

Two experimental areas were selected in pine forests of the Province of Albacete, Castilla—La Mancha Region, Central Eastern Spain. The first area (Sierra de las Quebradas, municipality of Liétor) was affected by a wildfire in July 2016. The second forest (municipality of Lezuza) was subjected to a prescribed fire in March 2016, to reduce fuel loading and thus the potential risk and severity of subsequent fires (Figure 1). Prior to wildfire, the soil cover of the forest was mainly composed of plants, litter and stones with variable composition.

Both study areas have a semiarid Mediterranean climate, BSk according to the Köppen–Geiger classification [42]. The average annual rainfall and medium annual temperature are 282 (Liétor) and 450 (Lezuza) mm and 13.5 (Liétor) and 16 (Lezuza) °C, respectively (Spanish National Meteorological Agency, 1950–2016). According to historical data (1990–2014) provided by the Spanish Meteorological Agency, the maximum precipitation is concentrated in October (44.5 mm) and the minimum in May (39.6 mm); from June to September a hot and dry period (air relative humidity below 50%) occurs. The mean minimum temperature of the coldest month is −0.9 °C and the mean maximum temperature of the hottest month is nearly 32 °C.

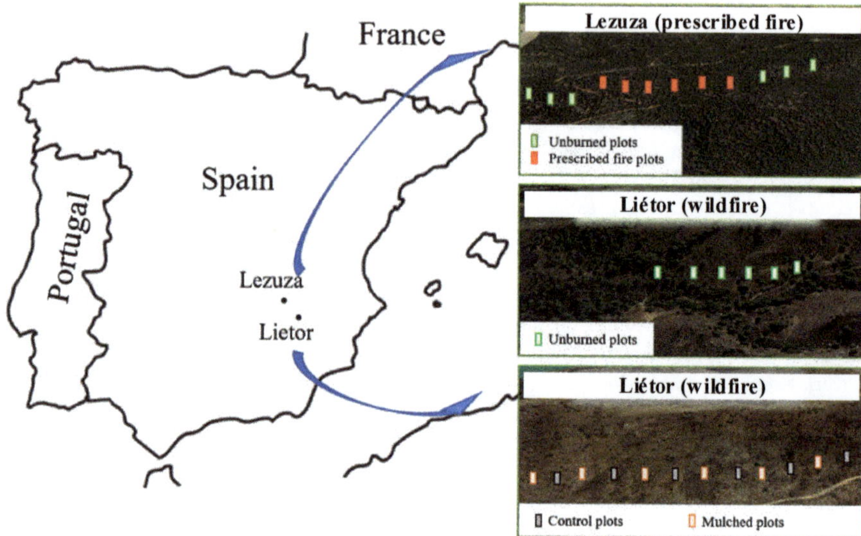

Figure 1. Location and layout of forest plots subject to prescribed fire and wildfire and monitored for hydrological observations (Lezuza and Liétor, Castilla La Mancha, Spain). Geographic coordinates and map source: Lezuza X: 557588E, Y: 4306475N; Liétor: X: 600081 E, Y: 4262798 N (unburned area); X: 598358 E, Y: 4264032 N (burned area); Google Earth, last access on 6/15/2019).

2.1.1. Wildfire-Affected Forest (Liétor)

The experimental area of Liétor is located at an elevation between 520 and 770 m a.s.l. with W-SW and N aspect and mean slope of 15–20% (Table 1). Soils are classified as Inceptisols and Aridisols with sandy-loam texture [43].

The wildfire burned about 830 ha of the forestland (mainly *Pinus halepensis* Mill). The mean tree density of this forest was between 500 and 650 trees per hectare with height between 7 and 14 m. *Rosmarinus officinalis* L., *Brachypodium retusum* (Pers.) Beauv., *Cistus clusii* Dunal, *Lavandula latifolia* Medik., *Thymus vulgaris* L., *Helichrysum stoechas* (L.) Moench, *Macrochloa tenacissima* (L.) Kunth, *Quercus coccifera* L. and *Plantago albicans* L. are the shrub or herbaceous species of the forest. The wildfire, classified as high-severity fire by the local forest managers according to the methodology proposed by Vega et al. [44], determined a tree mortality of 100% (Table 1). Forest floor was about 3–5 cm deep. This forest floor, as happened also in Lezuza, was blanketed with decaying *Pinus halepensis* M. needles and twigs and other wood debris such as cones or branches coming from trees. In both sites, the forest floor was blanketed with decaying *Pinus halepensis* M. needles and twigs and other wood debris such as cones or branches coming from trees in both sites, Lezuza and Liétor. More information related to this suggestion is provided in Table 1.

In September 2016 (two months after the wildfire), a mulching treatment was carried out in some areas of the burned site. Barley straw was spread manually on soil at a depth of 3 cm and a rate of 0.2 kg m^{-2} (dry weight), following the dose proposed by different authors for forests of Northern Spain, to achieve a burned soil cover over 80% [45].

Table 1. Main characteristics of forests and plots subject to prescribed fire and wildfire (Lezuza and Liétor, Castilla La Mancha, Spain).

Characteristics		Prescribed Fire (Lezuza)		Wildfire (Liétor)		
		Soil Condition				
		Control	Burned	Control	Burned	Burned and Mulched
Number of plots		6	6	6	6	6
Plot area (m^2)		8			200	
Elevation (m)		1010–1040			520–770	
Slope (%)		15 (±4.4)	14.5 (±2.6)		15–20	
Aspect		N	N-NE		W-SW and N	
Tree density (n ha^{-1})		477 (±33)	529 (±60)		500–650	
Tree height (m)		8–13			7–14	
Tree diameter (cm)		27 (±7)	32 (±6)		25–35	
Tree canopy cover (%)		75–85			60–70	
Tree species composition		*P. pinaster* Ait., *P. halepensis* M.	*P. pinaster* Ait., *P. halepensis* M., *Q. ilex* L.	*Pinus halepensis* Mill		
Shrub/herb species		*Quercus coccifera* L., *Brachypodium retusum* P., *Thymus vulgaris* L., *Dactylis glomerata* L.	*Quercus coccifera* L., *Brachypodium retusum* P., *Thymus vulgaris* L., *Sanguisorba minor* S.	*Rosmarinus officinalis* L., *Brachypodium retusum* (Pers.) Beauv., *Cistus clusii* Dunal, *Lavandula latifolia* Medik., *Thymus vulgaris* L., *Helichrysum stoechas* (L.) Moench, *Macrochloa tenacissima* (L.) Kunth, *Quercus coccifera* L. and *Plantago albicans* L.		
Canopy consumed by fire (%)		-	18 (±5)	-	90 (±15)	
Shrub/herb cover (%)	Pre-fire	56 (±9)	51 (±7)		65 (±9)	
	Post-fire	59 (±5)	10 (±3)		0 (±0)	
Litter (%)	Pre-fire	40 (±8)	44 (±7)		10 (±5)	
	Post-fire	38 (±9)	79 (±6)		5 (±2)	
Bare soil (%)	Pre-fire	4 (±1)	5 (±1)		25 (±8)	
	Post-fire	3 (±1)	11 (±3)		90 (±12)	

2.1.2. Forest Subjected to Prescribed Fire (Lezuza)

The forest area was selected in a relatively hilly area at an elevation from about 1010 to 1040 m a.s.l., with a 50-year old mixed plantation of *Pinus halepensis* and *Pinus pinaster*. The mean slope is around of 15% and the aspect is N-NE. Soils are classified as Alfisols with Xeralf Rhodoxeralf horizon with clay texture [43] (Table 1).

The tree density of this forest area was about 500 trees per hectare with a mean height of 6.40 m. The understory was dominated by *Quercus faginea* Lam. L., *Quercus ilex* subsp. ballota, *Quercus coccifera* L., *Juniperus oxycedrus*, *Brachypodium retusum* P. and *Thymus* sp. (Table 1). Forest floor depth was about 5–7 cm.

2.2. Description of Experimental Plots and Measurement of Runoff Volume

In the two selected forests, experimental plots were installed in unburned (control) and burned areas. Plots were randomly distributed in the experimental site in areas with the same morphological and ecological characteristics to ensure comparability.

More specifically, in the wildfire-affected forest of Liétor, eighteen rectangular plots (20-m long × 10-m wide) were installed at a distance between 200 and 500 m. Six plots were located in the forest outside of the burned site and assumed as a control. Twelve plots were instead located in the burned area, of which six were not treated, while six plots were mulched. In the forest subjected to the prescribed fire in Lezuza, twelve plots (4-m long and 2-m wide) were isolated, of which six were located in the unburned area and the other six in the burned site. The prescribed fire was carried out under controlled air conditions in the forests (wind speed of 14 km/h, air temperature of 14 °C and relative humidity of 63%), which are reference values for applying the prescribed fire as a forest protection measure. The upper and side borders of all plots in both areas were hydraulically isolated by geotextile fabric pounded into the ground, to prevent external water inputs. At the plot bottom, runoff was collected using a metal fence conveying the water into a pipe, which discharged to a plastic tank of 25 (Liétor) or 50 (Lezuza) liters. In these plots, immediately the runoff volumes were measured after each rainfall event throughout an observation period of about one year (Table 2).

Table 2. Main characteristics of precipitation events in plots subject to prescribed fire and wildfire (Lezuza and Liétor, Castilla La Mancha, Spain).

Event	Date	Days after Fire	Rainfall Height (mm)	Maximum Intensity (mm h^{-1})
		Prescribed fire (Lezuza)		
1	4 Apr 2016	5	20.0	8.8
2	6 May 2016	37	20.1	8.4
3	18 May 2016	49	10.2	4.3
4	12 Oct 2016	196	21.1	8.8
5	19 Oct 2016	203	27.4	5.3
6	8 Nov 2016	223	17.0	5.6
7	2 Dec 2016	247	52.4	4.2
8	23 Dec 2016	268	59.6	11.6
9	11 Feb 2017	318	38.2	6.3
10	4 Apr 2017	377	20.2	5.7
11	28 Apr 2017	394	28.2	6.8
		Wildfire (Liétor)		
1	21 Oct 2016	98	40.0	3.99
2	24 Nov 2016	129	41.0	1.48
3	8 Dec 2016	146	59.0	0.98
4	21 Dec 2016	159	93.8	2.1
5	30 Jan 2017	199	28.0	0.84
6	22 Feb 2017	222	16.8	1.14
7	8 Mar 2017	236	11.6	1.78
8	20 Mar 2017	248	102.6	16.2
9	12 May 2017	301	20.7	3.77

A weather station (WatchDog 2000 Series model) with a tipping bucket rain gauge was placed in each study area to measure total daily precipitation, storm duration, air temperature and rain intensity during the study period. In the hourly rainfall series of the experimental database, two consecutive events were considered separate, if no rainfall was recorded for 6 h or more [46,47]. The mean rainfall intensity was the total rainfall divided by the storm duration. The main variables characterizing the observed events are reported in Table 2.

2.3. Outlines on the SCS-CN Model

This model, proposed by the Soil Conservation Service of the United States Department of Agriculture in 1972, hypothesizes that:

$$\frac{V}{P_n} = \frac{W}{S} \tag{1}$$

being:

- V = runoff volume (mm);
- P_n = net precipitation (mm);
- W = water volume stored into the soil (mm);
- S = maximum water storage capacity of soil (mm).

P_n is the difference between the total precipitation (P) and the initial losses (I_a, as the water storage in the soil dips, interception, infiltration and evapotranspiration) prior to surface runoff. I_a is assumed to be proportional to S through a coefficient λ:

$$I_a = \lambda S \tag{2}$$

S is given by:

$$S = 25.4 \cdot \left(\frac{1000}{CN} - 10 \right) \tag{3}$$

when the parameter CN is the so called "curve number". The CN can be considered as the soil's aptitude to produce runoff and is a function of the hydrological properties and conditions of soil, and land use. The CN varies between 0 and 100 (0 means that the soil does not produces runoff, 100 means that all the precipitation turns into surface runoff and then the hydrological losses are zero).

According to this model, the runoff volume V is:

$$V = \frac{(P - \lambda S)^2}{P + (1 - \lambda)S} \tag{4}$$

To estimate CN in agroforest areas, the soil hydrological class, vegetation cover, hydrological condition (good, medium and poor) and cultivation practice and the antecedent moisture condition (AMC) of the soil must be determined.

The soil hydrological class (A to D) is related to the soil's capability to produce runoff, on its turn due to the soil infiltration capacity. A low runoff production capability corresponds to the A soil hydrological class, while the highest runoff capability is typical of less permeable soils D.

The actual AMC of the soil subject to a rainfall/runoff event was estimated as a function of the total height of precipitation in the five days before the event in the two different conditions of crop dormancy or the growing season. On this regard, three AMCs are identified:

- AMC_I: dry condition and minimum surface runoff;
- AMC_{II}: average condition and surface runoff;
- AMC_{III}: wet condition and maximum surface runoff.

The SCS guidelines make the CN values available in tables for a soil of given hydrological class and condition, vegetal cover, cultivation practice and average AMC (AMC_{II}). The values of CNs related to AMC_I (CN_I) or AMC_{III} (CN_{III}) can be calculated through the following equations:

$$CN_I = \frac{4.2CN_{II}}{10 - 0.058CN_{II}} \tag{5}$$

$$CN_{III} = \frac{23CN_{II}}{10 + 0.13CN_{II}} \tag{6}$$

The parameter AMC takes into account the influence of the soil water content on the hydrological response of the soil to the rainstorm and distinguishes "dry" (AMC_I), "average" (AMC_{II}) and "wet" (AMC_{III}) conditions depending on the total rainfall height of the five days before the event.

2.4. Model Implementation

2.4.1. Linear Regression between Rainfall and Runoff

A linear regression model was established between the surface runoff volume (dependent variable) and the rainfall height (independent variable) for each event, as follows:

$$V = aP \tag{7}$$

where:

- V = runoff volume (mm);
- P = total precipitation (mm);
- a = slope (-).

The intercept of this linear equation was forced to zero, in order to avoid runoff without any precipitation.

2.4.2. SCS-CN Model

The SCS-CN model was first applied considering the "default" input parameters, that is, the values of λ and CN derived from the SCS guidelines for woods (control plots) or pasture (burned plots) for the soil hydrologic group A of the experimental soils and AMC "I" for all the modeled events (since no or vey low precipitation was recorded in the antecedent five days). However, the runoff prediction capacity was totally unsatisfactory using default CNs, since very large errors between predictions and observations were achieved. Therefore, the SCS-CN model was adjusted by manual trials tuning both λ and CN parameters until the maximum coefficient of efficiency E (see below) was achieved using optimal λ and CN (Table 3).

Table 3. Optimal values of the Soil Conservation Service-curve number (SCS-CN) model parameters used for runoff predictions in plots subject to prescribed fire and wildfire (Lezuza and Liétor, Castilla La Mancha, Spain).

Input Parameter	Soil Condition				
	Prescribed Fire (Lezuza)		**Wildfire (Liétor)**		
	Unburned	Burned	Unburned	Burned	Burned and Mulched
Soil hydrologic class			A		
λ			0.0001		
CN	15	16	0.25	3 (27) *	3 (22) *
AMC			I		

Note: * indicates the CN value of the first modeled event.

2.5. Evaluation of Model Prediction Accuracy

The runoff simulations of the linear regression equation and SCS-CN model were analyzed for "goodness-of-fit" with the corresponding observations. More specifically, the observed and simulated runoff volumes were visually compared in scatterplots. Then, the main statistics and the indexes of goodness-of-fit commonly used in literature (e.g., [27,48–50]) were adopted (Table 4): (i) the maximum, minimum, mean and standard deviation of both the observed and simulated values; (ii) the coefficient of determination (r^2); (iii) the coefficient of efficiency of Nash and Sutcliffe [51] (E); (iv) the root mean square error (RMSE) and (v) the coefficient of residual mass (CRM, also known as "percent bias", PBIAS). Table 4 reports the equations and the range of variability of these indexes [52–55]. Generally speaking, these indexes are based on the analysis of the errors between simulations and predictions of the modeled hydrological variables.

Table 4. Indexes and related equations and range of variability to evaluate the runoff prediction capacity of the linear regression and curve number models in forest plots subject to prescribed fire and wildfire (Lezuza and Liétor, Castilla La Mancha, Spain).

Index	Equation	Range of Variability	Acceptance Limit and Notes
Coefficient of determination (r^2)	$r^2 = \left[\dfrac{\sum_{i=1}^{n} (O_i - \overline{O})(P_i - \overline{P})}{\sqrt{\sum_{i=1}^{n} (O_i - \overline{O})^2} \sqrt{\sum_{i=1}^{n} (P_i - \overline{P})^2}} \right]^2$	0 to 1	>0.5 [56–58]
Coefficient of efficiency (E, Nash and Sutcliffe [51])	$E = 1 - \dfrac{\sum_{i=1}^{n} (O_i - P_i)^2}{\sum_{i=1}^{n} (O_i - \overline{O})^2}$	$-\infty$ to 1	"Good" model accuracy if $E \geq 0.75$, "satisfactory" if $0.36 \leq E \leq 0.75$ and "unsatisfactory" if $E \leq 0.36$ [55]
Root mean square error (RMSE)	$RMSE = \sqrt{\dfrac{\sum_{i=1}^{n} (P_i - O_i)^2}{n}}$	0 to ∞	<0.5 of observed standard deviation [59]
Coefficient of residual mass (CRM or PBIAS, Loague and Green [50])	$CRM = \dfrac{\sum_{i=1}^{n} O_i - \sum_{i=1}^{n} P_i}{\sum_{i=1}^{n} O_i}$	$-\infty$ to ∞	<0.25 [54] CRM < 0 indicates model underestimation CRM > 0 indicates model overestimation [60]

Notes: n = number of observations; O_i, P_i = observed and predicted values at the time step i; $\overline{O}$ = mean of observed values.

3. Results and Discussion

3.1. Hydrological Characterization

3.1.1. Wildfire-Affected Forest (Liétor)

During the observation period, only nine events (total rainfall of 413 mm) produced surface runoff. For these events, precipitation height and mean intensity were in the range 11.6–93.8 mm and 0.98–28.0 mm/h, respectively. Expectedly, in the burned plots the runoff (on average 0.60 mm with a maximum value of 2.20 mm) was higher compared to the unburned soils (average of 0.03 mm and maximum of 0.08 mm). This may be due to the reduced infiltration and some combination of sealing, soil water repellency, loss of surface cover and decrease in soil aggregate stability, for the loss of organic matter [61].

In the mulched soil the mean and maximum runoff was 0.53 and 1.65 mm, respectively (Figure 2). These volumes were lower compared to the runoff generated in the burned plots. This shows the effectiveness of mulching as post-fire management technique to reduce the runoff generation capacity of the burned soils. These results confirm several other studies about the efficacy of mulch application, in order to control the hydrological response of soil after wildfires (e.g., [8,62–66]).

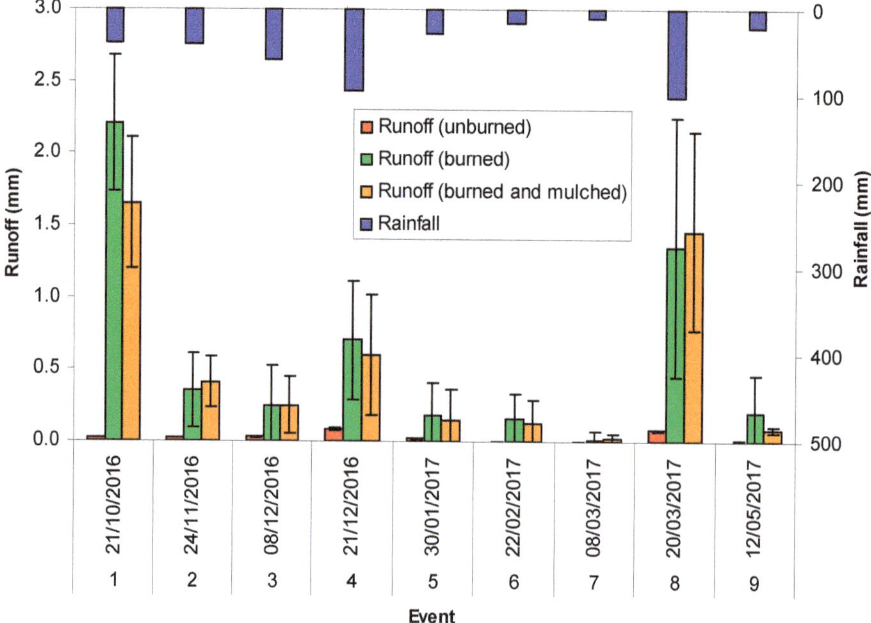

Figure 2. Rainfall and runoff volumes observed in forest plots subject to prescribed fire (Lezuza, Castilla La Mancha, Spain). Vertical bars are the standard deviations.

A sudden increase in the runoff generation capacity was evident in the first event after the fire (21 October 2016), presumably due to the ash release (that sealed the soil) and changes in the physicochemical properties (as the depletion in the organic matter content, which reduces the aggregate stability of the soil) [28,67]. A temporal reduction in runoff generation was found in burned soils (treated or not). This indicates a decrease in the hydrological response over time since the fire, also noticed by several authors in the early storms immediately after wildfire (e.g., [68–70]). The higher runoff is due to both the changes in soil hydrological properties and to the reduction of the vegetal cover after fire. As a matter of fact, the development of a water-repellent layer (also due to the ash released by fire) over the soil surface and the destruction of soil aggregates reduce water infiltration and thus increase runoff [71,72]. Over time, the shrub and herb vegetation quickly recovery, which decreases the runoff generation on the soil left bare by wildfire [73].

3.1.2. Forest Subjected to Prescribed Fire (Lezuza)

Sixteen storms (totaling 368 mm) produced runoff. The mean runoff from these rainfall events was 0.39 mm and the maximum was 0.69 mm. In the burned plots, the mean and maximum runoff volumes recorded were 0.40 and 0.75 mm, respectively (Figure 3). For few events, runoff from burned soils was lower compared to the control plots, while, for the majority of the monitored storms, the soil subjected to prescribed fire produced noticeably more runoff compared to the unburned plots. This waiving soil response to storms confirms the low impacts of low-intensity fires on the hydrological response of soils, already observed by several authors (e.g., [28,74]). This means that the prescribed fire has a limited potential to change the soil properties that drives the hydrological behavior, such as the repellency and infiltrability. However, as observed for the wildfire, attention should be paid to the first rainfall events occurring immediately after fire, when the removal of almost all the vegetal cover (including the litter) may leave the soil bare and thus exposed to rainfall erosivity.

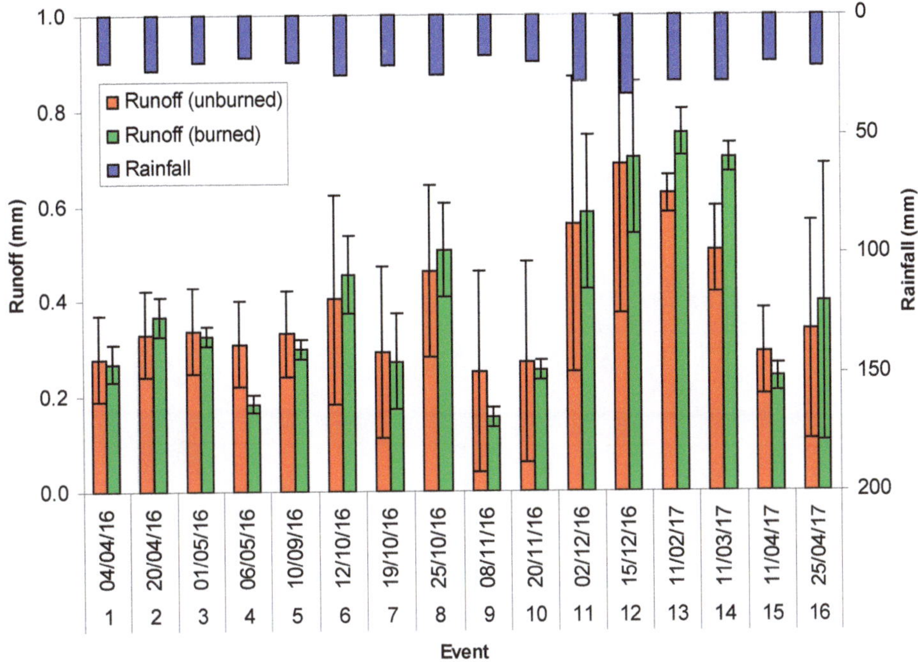

Figure 3. Rainfall and runoff volumes observed in forest plots subject to wildfire (Liétor, Castilla La Mancha, Spain). Vertical bars are the standard deviations.

3.2. Hydrological Modeling

3.2.1. Linear Regression

The simulation of runoff volumes gave satisfactory results for the unburned soil both in Lezuza and Liétor, as shown by the values of E (close or over 0.75) and r^2 (>0.62) indexes and the closeness between predictions and observations (mean error of 10-15%). RMSE values (0.01, Lezuza, and 0.07, Liétor) were under the limit of acceptance of Table 4 (half std. dev.) only for the runoff measured in Lezuza (0.165), but not in Liétor (0.005). In general, the linear models tended to overestimate the runoff volumes in unburned and burned and mulched plots (CRM < 0), while underestimating the runoff in burned plots (CRM > 0; Figure 4). In the latter condition, the maximum runoff values were noticeably underestimated (difference between 25 and 40%; Table 5).

The performance of the linear regression Equation (7) was only acceptable but not satisfactory in burned soils (with or without treatment), because E (between 0.52, soil burned by wildfire, and 0.62, soil burned by wildfire and then mulched) was lower than the suggested limit of Table 4 and the differences between the maximum values of observations and predictions were over 20%; only the mean values were close each others (error <12%; Table 5 and Figure 5a,b); the values of RMSE, which was acceptable in Lezuza (0.12 vs. a limit of 0.15), were 0.62 (burned and untreated plots) and 0.46 (burned and mulched soils) and therefore were over the acceptance limit (Table 4). This limited performance is mainly due to inaccurate prediction of the most intense rainfall–runoff event (21 October 2016), immediately following the fire. Moreover, for the burned soils, the RMSE values were higher than 50% of the standard deviation of observed runoff and thus not satisfactory; for soils burned by wildfire, also the coefficient of determination was poor (r^2 < 0.39). This unsatisfactory model performance is also visually shown by the large scattering of the simulations around the regression line (Figure 5a,b), which highlights a particular prediction inaccuracy for the first rainfall event (21 October 2016 recorded in Liétor). This inaccuracy is due to the soil changes induced by the wild fire (e.g., water repellency, decreases in infiltration and interception), which alters its hydrological response to rainstorms, but disappear some weeks or a few months after the fire [7–9,26]. Moreover, the linear equation was not able to simulate the variability of the hydrological processes with the precipitation, since the same runoff was observed for the same precipitation. This means that linear regressions are not able to simulate with reliability the surface runoff produced in burned plots, although these models may give an indication at least of the magnitude of the hydrological response of soils under different precipitation input and conditions.

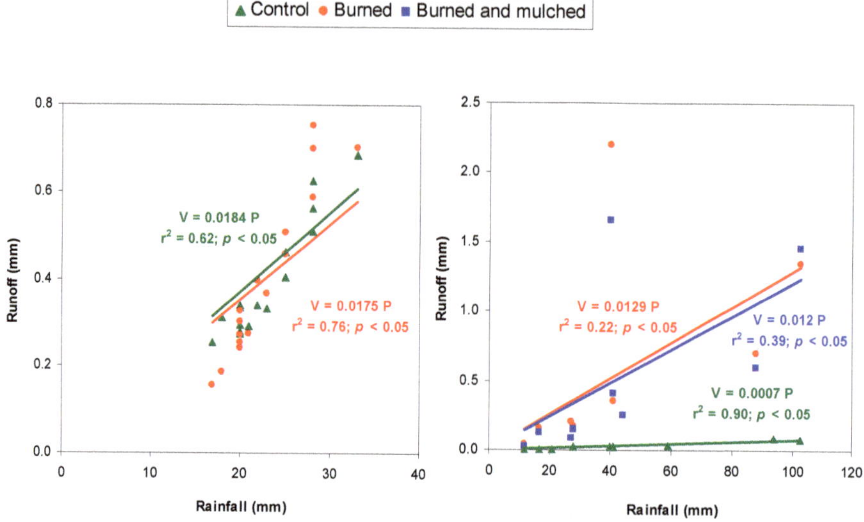

Figure 4. Linear regressions between observed rainfall and runoff in plots subject to prescribed fire and wildfire (Lezuza, **Left**, and Liétor, **Right**; Castilla La Mancha, Spain). V is the runoff volume and P is the rainfall height, while r2 and p are the coefficient of determination and the significance level, respectively.

Table 5. Statistics and indexes to evaluate the runoff prediction capacity of linear regression models in forest plots subject to prescribed fire and wildfire (Lezuza and Liétor, Castilla La Mancha, Spain).

Runoff Volume	Mean	Standard	Minimum	Maximum	r^2	E	CRM	RMSE
			Prescribed fire (Lezuza)					
			Control					
Observed	0.39	0.25	0.14	0.69	-	-	-	-
Simulated	0.42	0.31	0.08	0.61	0.62	0.73	−0.08	0.07
			Burned					
Observed	0.40	0.16	0.19	0.75	-	-	-	-
Simulated	0.40	0.30	0.08	0.58	0.75	0.60	0.01	0.12
			Wildfire (Liétor)					
			Control					
Observed	0.03	0.00	0.03	0.08	-	-	-	-
Simulated	0.03	0.01	0.02	0.07	0.90	0.89	−0.07	0.01
			Burned					
Observed	0.60	0.04	0.72	2.20	-	-	-	-
Simulated	0.59	0.15	0.43	1.32	0.22	0.52	0.02	0.62
			Burned and mulched					
Observed	0.49	0.01	0.62	1.66	-	-	-	-
Simulated	0.55	0.14	0.40	1.23	0.39	0.62	−0.11	0.46

Notes: r^2 = coefficient of determination; E = coefficient of efficiency; CRM = coefficient of residual mass; RMSE = root mean square error.

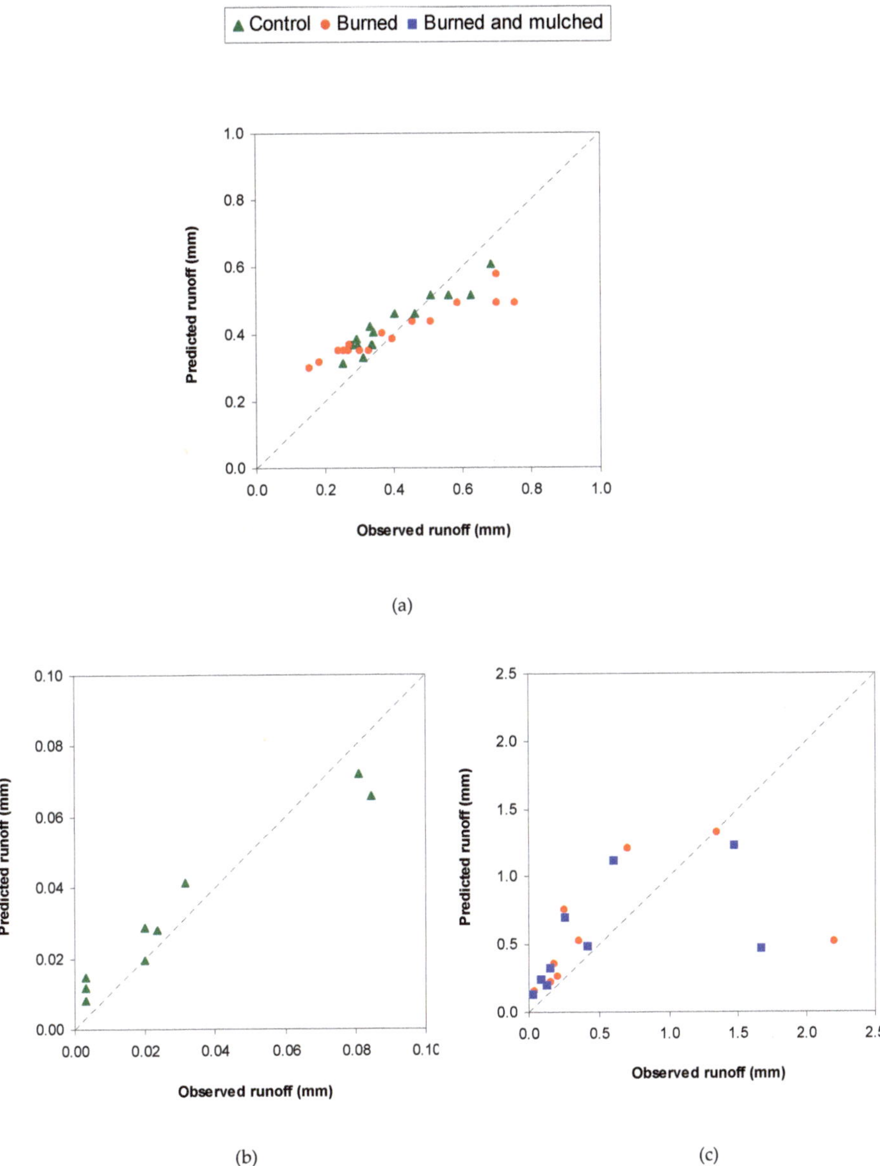

Figure 5. Scatterplots of runoff observations vs. predictions using linear regressions in plots subject to prescribed fire (Lezuza, **a**) and wildfire (Liétor, control soils, **b**, and burned soils **c**) (Castilla La Mancha, Spain).

3.2.2. SCS-CN Model

The predictions of runoff volume became more accurate for the majority of fire (in terms of severity) and soil conditions (Figure 6a,b). Compared to the linear regressions, the runoff predictions improved in the unburned plots of Lezuza (E = 0.87 and r^2 = 0.92), but slightly worsened in all the plots of Liétor (E = 0.88 and r^2 = 0.95; Table 6).

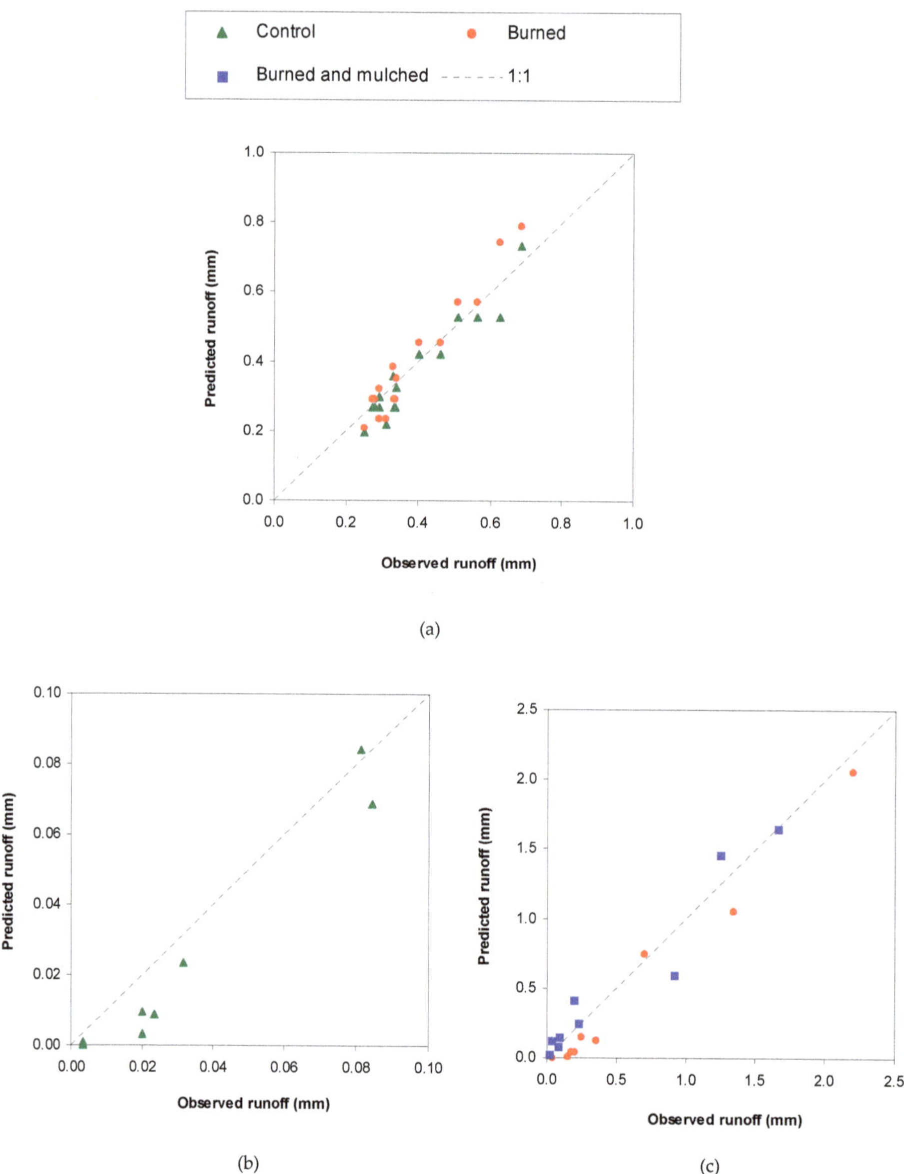

(a)

(b)

(c)

Figure 6. Scatterplots of runoff observations vs. predictions using the curve number model in plots subject to prescribed fire (Lezuza, **a**), and wildfire (Liétor, control soils, **b**, and burned soils, **c**) (Castilla La Mancha, Spain).

Conversely, the runoff was predicted with greater accuracy in burned soils (with or without treatment), as shown by E > 0.80—with peaks of 0.96-0.97 after wildfire—and r^2 > 0.94, these indicators being noticeably over the acceptance limit (Table 4). The RMSE values were always lower than 50% of the observed standard deviation (Table 6). In general, the SCS-CN model always showed a runoff overestimation (see CRM < 0). It should be noticed that the mean runoff values were predicted with less accuracy compared to the linear regression equations (differences between predictions and observations over 20% in some cases), but the SCS-CN model was much more reliable in predicting the maximum runoff volumes (differences lower than 7% for soils burned by wildfire and 15% for plots subjected to prescribed fire; Table 6). This indicates that, when accurate predictions of runoff are required to control the flooding risk (linked to the highest runoff volumes), the SCS-CN should be preferred to the simpler linear regressions.

Table 6. Statistics and indexes to evaluate the runoff prediction capacity of the curve number model in forest plots subject to prescribed fire and wildfire (Lezuza and Liétor, Castilla La Mancha, Spain).

Runoff Volume	Mean	Standard	Minimum	Maximum	r^2	E	CRM	RMSE
Prescribed fire (Lezuza)								
Control								
Observed	0.39	0.14	0.25	0.69	-	-	-	-
Simulated	0.37	0.15	0.20	0.73	0.92	0.87	0.06	0.05
Burned								
Observed	0.39	0.14	0.25	0.69	-	-	-	-
Simulated	0.41	0.18	0.21	0.79	0.95	0.81	-0.03	0.06
Wildfire (Liétor)								
Control								
Observed	0.03	0.03	0.003	0.08	-	-	-	-
Simulated	0.02	0.03	0.000	0.08	0.95	0.88	0.26	0.01
Burned								
Observed	0.60	0.72	0.04	2.20	-	-	-	-
Simulated	0.47	0.70	0.00	2.06	0.98	0.97	0.22	0.16
Burned and mulched								
Observed	0.49	0.62	0.01	1.66	-	-	-	-
Simulated	0.53	0.61	0.02	1.65	0.94	0.96	-0.07	0.15

Notes: r^2 = coefficient of determination; E = coefficient of efficiency; CRM = coefficient of residual mass; RMSE = root mean square error.

Some additional considerations about SCS-CN model application in the experimental conditions should be made.

First, very low values of CN and λ were provided in this study as input to the model, in order to predict with accuracy runoff after the two fire-severity conditions. This means that the water losses during and immediately after the rainfall (reflected by I_a and S, such as water storage in the soil dips, interception, infiltration and evapotranspiration) are very high and the storms produce very small runoff volumes. In more detail, the small CN simulates a large water storage capacity (S) of soil through the infiltration process and λ must be decreased even by three orders of magnitude to simulate the very low initial water losses, due to interception and evapo-transpiration.

Second, both for wildfire and prescribed fire, unrealistic input parameters are required to simulate such a minimal runoff generation capacity of these soils. As a matter of fact, values of 15–16 (after wildfire) or even 0.25–3 (after prescribed fire) for CN and 0.0001 for λ against common values over 30 for CN and 0.2 for λ are needed to fit the runoff predictions to the corresponding observations. This should be taken into account when the SCS-CN model must be implemented in soils having a small hydrological response.

Third, a unique CN value as input for the SCS-CN model is not able to reproduce the increase in runoff immediately after wildfire. The worsening of the hydrological response of the burned soil both after wildfire and prescribed fire has been shown by a number of studies (e.g., [9]) and particularly in Mediterranean forests (e.g., Keizer et al. [8], in a Portuguese eucalypt forest; Lucas-Borja et al. [7,62], in pine forests of Central Spain). This increase is mainly due to soil water repellency and vegetation cover removal due to fires, but these effects disappear some months after fire. In order to simulate the hydrological effects of a repellent and almost bare soil, it is necessary to increase the CNs in this so-called "window-of-disturbance" [15,75] up to values that may be noticeably high. Soil mulching "smoothes" the increase in the soil hydrological response and this requires a lower increase in CN values.

4. Conclusions

This study evaluated the runoff prediction capacity of the SCS-CN model in comparison with linear regression equations after storms in two pine forests of Central-Eastern Spain affected by wildfire (with or without a rehabilitation treatment using straw mulching) or prescribed fire.

The simulation of runoff volumes by the linear regression gave satisfactory results only for the unburned soils. Conversely, for the burned plots, the linear regressions failed in simulating the runoff with reliability. The SCS-CN model was instead accurate to predict the runoff volume particularly in burned soils, also when mulching was applied. Although the mean runoff was predicted with less accuracy compared to the linear equations, the model performance was very satisfactory in predicting the maximum volumes. Moreover, all the soil conditions (unburned, burned and burned and mulched) were simulated with reliability. To reproduce the peculiar hydrology of the experimental areas, very low values of CNs and initial abstraction were required, which may appear unrealistic; moreover, the post-fire hydrological window-of-disturbance could be reproduced only by increasing the CN for the storms occurring few months after wildfire.

The performances of the two tested models indicate that, in Mediterranean forests subject to the fire risk, the use of simple linear equations is suggested for predicting runoff generated by relatively low storms, while the SCS-CN model is more reliable and therefore advisable when runoff predictions are needed to control the flooding risk.

Overall, the study has confirmed the viability of the SCS-CN method to reproduce the complex hydrological response of unburned and burned (and treated or not) soils of Mediterranean forests. Although this assumption is limited to the experimental conditions, the results are encouraging towards larger applications of this method in other climatic and geomorphologic conditions. However, further modeling studies are needed, in order to explore the runoff prediction capacity of the model in fire-affected forests with different ecological and soil characteristics. These studies should also be enlarged from the plot to the watershed scale, using more complex hydrological models based on the SCS-CN method. Once validated in a wide range of environmental contexts, the use of these models may support the land managers to control runoff and erosion in mountain forests that are prone to both the wildfire and hydrogeological risks.

Author Contributions: Conceptualization, D.A.Z.; methodology, A.L., D.D., P.D., P.A.P.-A. and B.G.C.; formal analysis, G.B. and M.E.L.-B.; investigation, A.L., D.D., P.D., P.A.P.-A. and B.G.C.; resources, M.E.L.-B.; data curation, A.L., D.D., P.D., P.A.P.-A. and B.G.C.; writing—Original draft preparation, D.A.Z. and M.E.L.-B.; writing—review and editing, D.A.Z., G.B. and M.E.L.-B.; supervision, D.A.Z., G.B. and M.E.L.-B. All authors have read and agreed to the published version of the manuscript.

Funding: This research received no external funding.

Acknowledgments: Pedro Antonio Plaza Álvarez was supported by a predoctoral fellowship from the Spanish Ministry of Education, Culture and Sport (FPU16/03296).

Conflicts of Interest: The authors declare no conflict of interest.

References

1. Hou, Y.; Zhang, M.; Liu, S.; Sun, P.; Yin, L.; Yang, T.; Yide, L.; Qiang, L.; Wei, X. The hydrological impact of extreme weather-induced forest disturbances in a tropical experimental watershed in south China. *Forests* **2018**, *9*, 734. [CrossRef]

2. Aryal, Y.; Zhu, J. Effect of watershed disturbance on seasonal hydrological drought: An improved double mass curve (IDMC) technique. *J. Hydrol.* **2020**, *585*, 124746. [CrossRef]

3. Lindner, M.; Maroschek, M.; Netherer, S.; Kremer, A.; Barbati, A.; Garcia-Gonzalo, J.; Lexer, M.J. Climate change impacts, adaptive capacity, and vulnerability of European forest ecosystems. *For. Ecol. Manag.* **2010**, *259*, 698–709. [CrossRef]

4.	Flanagan, S.A.; Hurtt, G.C.; Fisk, J.P.; Sahajpal, R.; Hansen, M.C.; Dolan, K.A.; Sullivan, J.H.; Zhao, M. Potential Vegetation and Carbon Redistribution in Northern North America from Climate Change. *Climate* **2016**, *4*, 2. [CrossRef]

5.	Nunes, L.J.; Meireles, C.I.; Pinto Gomes, C.J.; Almeida Ribeiro, N.M. Forest Contribution to Climate Change Mitigation: Management Oriented to Carbon Capture and Storage. *Climate* **2020**, *8*, 21. [CrossRef]

6.	Niemeyer, R.J.; Bladon, K.D.; Woodsmith, R.D. Long term hydrologic recovery after wildfire and post-fire forest management in the interior Pacific Northwest. *Hydrol. Process.* **2020**, *34*, 1182–1197. [CrossRef]

7.	Lucas-Borja, M.E.; González-Romero, J.; Plaza-Álvarez, P.A.; Sagra, J.; Gómez, M.E.; Moya, D.; de Las Heras, J. The impact of straw mulching and salvage logging on post-fire runoff and soil erosion generation under Mediterranean climate conditions. *Sci. Total Environ.* **2019**, *654*, 441–451. [CrossRef]

8.	Keizer, J.J.; Silva, F.C.; Vieira, D.C.; González-Pelayo, O.; Campos, I.M.A.N.; Vieira, A.M.D.; Prats, S.A. The effectiveness of two contrasting mulch application rates to reduce post-fire erosion in a Portuguese eucalypt plantation. *Catena* **2018**, *169*, 21–30. [CrossRef]

9.	Wilson, C.; Kampf, S.K.; Wagenbrenner, J.W.; MacDonald, L.H. Rainfall thresholds for post-fire runoff and sediment delivery from plot to watershed scales. *For. Ecol. Manag.* **2018**, *430*, 346–356. [CrossRef]

10.	Kinoshita, A.M.; Chin, A.; Simon, G.L.; Briles, C.; Hogue, T.S.; O'Dowd, A.P.; Albornoz, A.U. Wildfire, water, and society: Toward integrative research in the "Anthropocene". *Anthropocene* **2016**, *16*, 16–27. [CrossRef]

11.	Smith, H.G.; Sheridan, G.J.; Lane, P.N.; Nyman, P.; Haydon, S. Wildfire effects on water quality in forest catchments: A review with implications for water supply. *J. Hydrol.* **2011**, *396*, 170–192. [CrossRef]

12.	Menéndez-Duarte, R.; Fernández, S.; Soto, J. The application of 137Cs to post-fire erosion in north-west Spain. *Geoderma* **2009**, *150*, 54–63. [CrossRef]

13.	Lopéz, R.; Batalla, R.J. Análisis del comportamiento hidrológico de la cuenca mediterránea de Arbúbices antes y después de un incendio forestal. III Congreso Forestal, Español, Granada, Spain. 2001. Available online: ile:///C:/Users/MDPI/AppData/Local/Temp/3CFE01-084.pdf (accessed on 16 November 2020).

14.	Shakesby, R.A. Post-wildfire soil erosion in the Mediterranean: Review and future research directions. *Earth Sci. Rev.* **2011**, *105*, 71–100. [CrossRef]

15.	Zavala, L.M.M.; de Celis Silvia, R.; López, A.J. How wildfires affect soil properties. A brief review. *Cuad. Investig. geográfica/Geogr. Res. Lett.* **2014**, *40*, 311–331. [CrossRef]

16.	Moody, J.A.; Shakesby, R.A.; Robichaud, P.R.; Cannon, S.H.; Martin, D.A. Current research issues related to post-wildfire runoff and erosion processes. *Earth Sci. Rev.* **2013**, *122*, 10–37. [CrossRef]

17.	Nunes, J.P.; BernardJannin, L.; Rodriguez Blanco, M.L.; Santos, J.M.; Coelho, C.D.O.A.; Keizer, J.J. Hydrological and erosion processes in terraced fields: Observations from a humid Mediterranean region in northern Portugal. *Land Degrad. Dev.* **2018**, *29*, 596–606. [CrossRef]

18.	Nunes, J.P.; Naranjo Quintanilla, P.; Santos, J.M.; Serpa, D.; Carvalho-Santos, C.; Rocha, J.; Keesstra, S.D. Afforestation, subsequent forest fires and provision of hydrological services: A model-based analysis for a Mediterranean mountainous catchment. *Land Degrad. Dev.* **2018**, *29*, 776–788. [CrossRef]

19.	Nunes, J.P.; Doerr, S.H.; Sheridan, G.; Neris, J.; Santín, C.; Emelko, M.B.; Silins, U.; Robichaud, P.R.; Elliot, W.J.; Keizer, J. Assessing water contamination risk from vegetation fires: Challenges, opportunities and a framework for progress. *Hydrol. Process.* **2018**, *32*, 687–1684. [CrossRef]

20.	Filianoti, P.; Gurnari, L.; Zema, D.A.; Bombino, G.; Sinagra, M.; Tucciarelli, T. An evaluation matrix to compare computer hydrological models for flood predictions. *Hydrology* **2020**, *7*, 42. [CrossRef]

21.	Lucas-Borja, M.E.; Zema, D.A.; Carrà, B.G.; Cerdà, A.; Plaza-Alvarez, P.A.; Cózar, J.S.; de las Heras, J. Short-term changes in infiltration between straw mulched and non-mulched soils after wildfire in Mediterranean forest ecosystems. *Ecol. Eng.* **2018**, *122*, 27–31. [CrossRef]

22.	IPCC. IPCC's 5th Assessment Report for Europe. 2013. Available online: http://www.ipcc.ch/pdf/assessment-report/ar5/wg2/WGIIAR5-Chap23_FINAL.pdf (accessed on 31 August 2018).

23.	Bedia, J.; Herrera, S.; Camia, A.; Moreno, J.M.; Gutiérrez, J.M. Forest fire danger projections in the Mediterranean using ENSEMBLES regional climate change scenarios. *Clim. Chang.* **2014**, *122*, 185. [CrossRef]

24.	Zema, D.A.; Lucas-Borja, M.E.; Fotia, L.; Rosaci, D.; Sarnè, G.M.; Zimbone, S.M. Predicting the hydrological response of a forest after wildfire and soil treatments using an Artificial Neural Network. *Comput. Electron. Agric.* **2020**, *170*, 105280. [CrossRef]

25. Zema, D.A.; Nunes, J.P.; Lucas-Borja, M.E. Improvement of seasonal runoff and soil loss predictions by the MMF (Morgan-Morgan-Finney) model after wildfire and soil treatment in Mediterranean forest ecosystems. *Catena* **2020**, *188*, 104415. [CrossRef]

26. Merritt, W.S.; Letcher, R.A.; Jakeman, A.J. A review of erosion and sediment transport models. *Environ. Model. Softw.* **2003**, *18*, 761–799. [CrossRef]

27. Zema, D.A.; Lucas-Borja, M.E.; Carrà, B.G.; Denisi, P.; Rodrigues, V.A.; Ranzini, M.; Soriano, F.C.; de Cicco, V.; Zimbone, S.M. Simulating the hydrological response of a small tropical forest watershed (Mata Atlantica, Brazil) by the AnnAGNPS model. *Sci. Total Environ.* **2018**, *636*, 737–750. [CrossRef]

28. Pereira, P.; Francos, M.; Brevik, E.C.; Ubeda, X.; Bogunovic, I. Post-fire soil management. *Curr. Opin. Environ. Sci. Health* **2018**, *5*, 26–32. [CrossRef]

29. Lucas-Borja, M.E.; Carrà, B.G.; Nunes, J.P.; Bernard-Jannin, L.; Zema, D.A.; Zimbone, S.M. Impacts of land-use and climate changes on surface runoff in a tropical forest watershed (Brazil). *Hydrol. Sci. J.* **2020**, *65*, 1956–1973. [CrossRef]

30. Mishra, S.K.; Singh, V.P. Another look at SCS-CN method. *J. Hydrol. Eng.* **1999**, *4*, 257–264. [CrossRef]

31. Mishra, S.K.; Singh, V.P. *Soil Conservation Service Curve NUMBER (SCS-CN) methodology (Vol. 42)*; Springer Science & Business Media: Berlin/Heidelberg, Germany, 2013; p. 277.

32. Suresh Babu, P.; Mishra, S.K. Improved SCS-CN–inspired model. *J. Hydrol. Eng.* **2012**, *17*, 1164–1172. [CrossRef]

33. Michel, C.; Andreassian, V.; Perrin, C. Soil conservation service curve number method: How to mend a wrong soil moisture accounting procedure. *Water Resour. Res.* **2005**, *41*, W02011. [CrossRef]

34. Jain, M.K.; Mishra, S.K.; Suresh Babu, P.; Venugopal, K. On the Ia–S relation of the SCS-CN method. *Hydrol. Res.* **2006**, *37*, 261–275. [CrossRef]

35. Garen, D.; Moore, D.S. Curve number hydrology in water quality modeling: Uses, abuses, and future directions. *J. Am. Water Resour. Assoc.* **2005**, *41*, 377–388. [CrossRef]

36. Ponce, V.M.; Hawkins, R.H. Runoff curve number: Has it reached maturity? *J. Hydrol. Eng.* **1996**, *1*, 11–19. [CrossRef]

37. Soulis, K.X. Estimation of SCS Curve Number variation following forest fires. *Hydrol. Sci. J.* **2018**, *63*, 1332–1346. [CrossRef]

38. Hawkins, R.H.; Greenberg, R.J. *WILDCAT4 Flow Model*; University of Arizona, School of Renewable Natural Resources: Tucson, AZ, USA, 1990.

39. Cerrelli, G.A. Fire hydro, a simplified method for predicting peak discharges to assist in the design of flood protection measures for western wildfires. In *Proceedings of the 2005 Watershed Management Conference—Managing Watersheds for Human and Natural Impacts: Engineering, Ecological, and Economic Challenges, Williamsburg, VA, USA, 19–22 July 2005*; Moglen, G.E., Ed.; American Society of Civil Engineers: Alexandria, VA, USA, 2005; pp. 935–941.

40. Candela, A.; Aronica, G.; Santoro, M. Effects of Forest Fires on Flood Frequency Curves in a Mediterranean Catchment/Effets d'incendies de forêt sur les courbes de fréquence de crue dans un bassin versant Méditerranéen. *Hydrol. Sci. J.* **2005**, *50*, 1–206. [CrossRef]

41. Nunes, J.P.; Catarina Simões Vieira, D.; Keizer, J.J. Comparing simple and complex approaches to simulate the impacts of soil water repellency on runoff and erosion in burnt Mediterranean forest slopes. In Proceedings of the European Geosciences Union, General Assembly, EGUGA, Vienna, Austria, 23 April 2017; p. 10815.

42. Kottek, M.; Grieser, J.; Beck, C.; Rudolf, B.; Rubel, F. World map of the Köppen-Geiger climate classification updated. *Meteorol. Z.* **2006**, *15*, 259–263. [CrossRef]

43. Soil Survey Staff. *Soil Taxonomy. Soil Taxonomy: A Basic System of Soil Classification for Making and Interpreting Soil Surveys. Agricultural Handbook 436*, 2nd ed.; Natural Resources Conservation Service, USDA: Washington, DC, USA, 1999; p. 869.

44. Vega, J.A.; Fontúrbel, T.; Merino, A.; Fernández, C.; Ferreiro, A.; Jiménez, E. Testing the ability of visual indicators of soil burn severity to reflect changes in soil chemical and microbial properties in pine forests and shrubland. *Plant Soil.* **2013**, *369*, 73–91. [CrossRef]

45. Vega, J.A.; Fernández, C.; Fonturbel, T.; González-Prieto, S.; Jiménez, E. Testing the effects of straw mulching and herb seeding on soil erosion after fire in a gorse shrubland. *Geoderma* **2014**, *223*, 79–87. [CrossRef]

46. Zema, D.A.; Labate, A.; Martino, D.; Zimbone, S.M. Comparing different infiltration methods of the HEC-HMS model: The case study of the Mésima Torrent (Southern Italy). *Land Degrad. Dev.* **2017**, *28*, 294–308. [CrossRef]

47. Wischmeier, W.H.; Smith, D.D. *Prediction Rainfall Erosion Losses*; Handbook, No. 537; USDA: Washington, DC, USA, 1978.

48. Willmott, C.J. Some comments on the evaluation of model performance. *Bull. Am. Meteorol. Soc.* **1982**, *63*, 1309–1313. [CrossRef]

49. Legates, D.R.; McCabe, G.J. Evaluating the use of "goodness of fit" measures in hydrologic and hydroclimatic model validation. *Water Resour. Res.* **1999**, *35*, 233–241. [CrossRef]

50. Loague, K.; Green, R.E. Statistical and graphical methods for evaluating solute transport models: Overview and application. *J. Contam. Hydrol.* **1991**, *7*, 51–73. [CrossRef]

51. Nash, J.E.; Sutcliffe, J.V. River flow forecasting through conceptual models: Part I. A discussion of principles. *J. Hydrol.* **1970**, *10*, 282–290. [CrossRef]

52. Zema, D.A.; Bingner, R.L.; Govers, G.; Licciardello, F.; Denisi, P.; Zimbone, S.M. Evaluation of runoff, peak flow and sediment yield for events simulated by the AnnAGNPS model in a Belgian agricultural watershed. *Land Degrad. Dev.* **2012**, *23*, 205–215. [CrossRef]

53. Krause, P.; Boyle, D.P.; Base, F. Comparison of different efficiency criteria for hydrological model assessment. *Adv. Geosci.* **2005**, *5*, 89–97. [CrossRef]

54. Moriasi, D.N.; Arnold, J.G.; Van Liew, M.W.; Bingner, R.L.; Harmel, R.D.; Veith, T.L. Model evaluation guidelines for systematic quantification of accuracy in watershed simulations. *Trans. ASABE* **2007**, *50*, 885–900. [CrossRef]

55. Van Liew, M.W.; Garbrecht, J. Hydrologic simulation of the Little Washita River experimental watershed using SWAT. *J. Am. Water Resour. Assoc.* **2003**, *39*, 413–426. [CrossRef]

56. Santhi, C.; Arnold, J.G.; Williams, J.R.; Dugas, W.A.; Srinivasan, R.; Hauck, L.M. Validation of the SWAT model on a large river basin with point and nonpoint sources. *J. Am. Water Resour. Assoc.* **2001**, *37*, 1169–1188. [CrossRef]

57. Van Liew, M.W.; Arnold, J.G.; Garbrecht, J.D. Hydrologic simulation on agri- cultural watersheds: Choosing between two models. *Trans. ASABE* **2003**, *46*, 1539–1551. [CrossRef]

58. Vieira, D.C.S.; Prats, S.A.; Nunes, J.P.; Shakesby, R.A.; Coelho, C.O.A.; Keizer, J.J. Modelling runoff and erosion, and their mitigation, in burned Portuguese forest using the revised Morgan-Morgan-Finney model. *For. Ecol. Manag.* **2014**, *314*, 150–165. [CrossRef]

59. Singh, J.; Knapp, H.V.; Demissie, M. Hydrologic modeling of the Iroquois River watershed using HSPF and SWAT. ISWS CR 2004-08. Champaign, Ill.: Illinois State Water Survey. 2004. Available online: http://www.sws.uiuc.edu/pubdoc/CR/ISWSCR2004-08.pdf (accessed on 14 February 2018).

60. Gupta, H.V.; Sorooshian, S.; Yapo, P.O. Status of automatic calibration for hydrologic models: Comparison with multilevel expert calibration. *J. Hydrol. Eng.* **1999**, *4*, 135–143. [CrossRef]

61. Neary, D.G.; Ryan, K.C.; DeBano, L.F. Wildland fire in ecosystems: Effects of fire on soils and water. Gen. Tech. Rep. RMRS-GTR-42-vol. 4. Ogden, UT: US Department of Agriculture, Forest Service, Rocky Mountain Research Station. 250 p. 42. 2005. Available online: https://www.fs.fed.us/rm/pubs/rmrs_gtr042_4.pdf (accessed on 16 November 2020).

62. Lucas-Borja, M.E.; Plaza-Álvarez, P.A.; Gonzalez-Romero, J.; Sagra, J.; Alfaro-Sánchez, R.; Zema, D.A.; de Las Heras, J. Short-term effects of prescribed burning in Mediterranean pine plantations on surface runoff, soil erosion and water quality of runoff. *Sci. Total Environ.* **2019**, *674*, 615–622. [CrossRef]

63. Prats, S.A.; González-Pelayo, Ó.; Silva, F.C.; Bokhorst, K.J.; Baartman, J.E.; Keizer, J.J. Post-fire soil erosion mitigation at the scale of swales using forest logging residues at a reduced application rate. *Earth Surf. Process. Landf.* **2019**, *44*, 2837–2848. [CrossRef]

64. Prats, S.A.; Malvar, M.C.; Coelho, C.O.A.; Wagenbrenner, J.W. Hydrologic and erosion responses to compaction and added surface cover in post-fire logged areas: Isolating splash, interrill and rill erosion. *J. Hydrol.* **2019**, *575*, 408–419. [CrossRef]

65. Lopes, A.R.; Prats, S.A.; Silva, F.C.; Keizer, J.J. Effects of ploughing and mulching on soil and organic matter losses after a wildfire in Central Portugal. *Cuad. Investig. Geográfica* **2020**, *46*, 303–318. [CrossRef]

66. Robichaud, P.R.; Lewis, S.A.; Brown, R.E.; Bone, E.D.; Brooks, E.S. Evaluating post-wildfire logging-slash cover treatment to reduce hillslope erosion after salvage logging using ground measurements and remote sensing. *Hydrol. Process.* **2020**, *34*, 4431–4445. [CrossRef]

Climate **2020**, *8*, 150

67. Keesstra, S.; Wittenberg, L.; Maroulis, J.; Sambalino, F.; Malkinson, D.; Cerdà, A.; Pereira, P. The influence of fire history, plant species and post-fire management on soil water repellency in a Mediterranean catchment: The Mount Carmel range, Israel. *Catena* **2017**, *149*, 857–866. [CrossRef]

68. de Dios Benavides-Solorio, J.; MacDonald, L.H. Measurement and prediction of post-fire erosion at the hillslope scale, Colorado Front Range. *Int. J. Wildland Fire* **2005**, *14*, 457–474. [CrossRef]

69. DeBano, L.F.; Debano, L.F.; Neary, D.G.; Ffolliott, P.F. *Fire Effects on Ecosystems*; John Wiley & Sons: Hoboken, NJ, USA, 1998.

70. MacDonald, L.H.; Sampson, R.; Brady, D.; Juarros, L.; Martin, D. Chapter 4. Predicting post-fire erosion and sedimentation risk on a landscape scale: A case study from Colorado. *J. Sustain. For.* **2000**, *11*, 57–87. [CrossRef]

71. DeBano, L.F.; Mann, L.D.; Hamilton, D.A. Translocation of hydrophobic substances into soil by burning organic litter 1. *Soil. Sci. Soc. Am. J.* **1970**, *34*, 130–133. [CrossRef]

72. Shakesby, R.A.; Doerr, S.H.; Walsh, R.P.D. The erosional impact of soil hydrophobicity: Current problems and future research directions. *J. Hydrol.* **2000**, *231*, 178–191. [CrossRef]

73. Díaz-Delgado, R.; Pons, X. Spatial patterns of forest fires in Catalonia (NE of Spain) along the period 1975–1995: Analysis of vegetation recovery after fire. *For. Ecol. Manag.* **2001**, *147*, 67–74. [CrossRef]

74. Keeley, J.E. Fire intensity, fire severity and burn severity: A brief review and suggested usage. *Int. J. Wildland Fire* **2009**, *18*, 116–126. [CrossRef]

75. Prosser, I.P.; Williams, L. The effect of wildfire on runoff and erosion in native Eucalyptus forest. *Hydrol. Process.* **1998**, *12*, 251–265. [CrossRef]

Publisher's Note: MDPI stays neutral with regard to jurisdictional claims in published maps and institutional affiliations.

Article

Development of a Parametric Regional Multivariate Statistical Weather Generator for Risk Assessment Studies in Areas with Limited Data Availability

Saddam Q. Waheed [1,2,*], Neil S. Grigg [1] and Jorge A. Ramirez [1,†]

[1] Department of Civil and Environmental Engineering, Colorado State University, 1372 Campus Delivery, Fort Collins, CO 80523-1372, USA; neil.grigg@colostate.edu (N.S.G.); Jorge.Ramirez@colostate.edu (J.A.R.)

[2] Iraqi Ministry of Water Resources, Planning and Follow up Directorate, Palestine Street, Baghdad, Iraq

* Correspondence: saddam.waheed@colostate.edu

† Deceased on March 28 2020; formerly, Professor, Dept. of Civil and Environmental Engineering.

Received: 10 July 2020; Accepted: 7 August 2020; Published: 11 August 2020

Abstract: Risk analysis of water resources systems can use statistical weather generators coupled with hydrologic models to examine scenarios of extreme events caused by climate change. These require multivariate, multi-site models that mimic the spatial, temporal, and cross correlations of observed data. This study developed a statistical weather generator to facilitate bottom-up approaches to assess the impact of climate change on water resources systems for cases of limited data. While existing weather generator models have impressive features, this study suggested a simple weather generator which is straightforward to implement and can employ any distribution function for variables such as precipitation or temperature. It is based on (1) a first-order, two-state Markov chain to simulate precipitation occurrences; (2) the use of Wilks' technique to produce correlated weather variables at multiple sites with the conservation of spatial, temporal, and cross correlations; (3) the capability to vary the statistical parameters of the weather variables. The model was applied to studies of the Diyala River basin in Iraq, which is a case with limited observed records. Results show that it exhibits high values (e.g., over 0.95) for the Nash–Sutcliffe and Kling–Gupta metric tests, preserves the statistical properties of the observed variables, and conserves the spatial, temporal, and cross correlations among the weather variables in the meteorological stations.

Keywords: statistical weather generator; stochastic process; Diyala River basin; Wilks' technique

---

1. Introduction

Climate change impacts are of increasing concern to hydrologists who assess risks in the management of water resources systems. Their models of climate scenarios for extreme events can be derived from global climate models (GCMs), stochastic-statistical weather generators (SWGs), or a combination. Although they have their own advantages, some argue that the GCM scenarios are inadequate and limit decision-making options because they represent only specific scenarios for climatic variability and have large uncertainties [1–6]. On the other hand, others think that SWGs can produce a wide range of scenarios to study system responses and provide more insights about the system performance under climate change [7–10]. The drawbacks of the SWGs are that they have a stochastic-basis and cannot provide future change insights. Therefore, the SWGs and GCMs have been linked to generate forecasting scenarios and to assign a probability of each SWG scenario by fitting a distribution to the GCM outcomes [11–13]. In this way, SWGs can then be used to generate probabilistic synthetic scenarios with the aid of the GCM information and which are statistically similar to observed data and used to investigate which climate states cause system failure [4,14–23]. Where historic records are limited, synthetic weather sequences based on SWGs are especially suitable [24].

Given the previous work, the main objective of this paper is to develop a SWG that can be used in a bottom-up approach to generate daily synthetic scenarios to evaluate the impacts of long-term climate change on system performance and suggest robust adaptations to cope with anticipated negative impacts that will be examined in a follow up study. Emphasis is placed on areas with low data availability, and the model is demonstrated for Diyala River basin in Iraq for the four historic weather variables (e.g., precipitation, maximum and minimum temperatures, and wind speed magnitude) with daily time steps from 1948 to 2006.

2. Literature Review

Generally, SWGs can be grouped into parametric, non-parametric, and semi-parametric methods. In the parametric method, the weather variables are assumed to fit one continuous probability distribution or two combined distributions. The parameters are usually estimated from historic observations [24–28]. In the non-parametric method, the weather variables are resampled from historic observations using techniques such as empirical distributions, neural networks, and maximum entropy bootstrap [29–32]. The semi-parametric method is a mixture between parametric and non-parametric methods.

Albeit other approaches have their advantages, the parametric SWG in the bottom-up approach is preferable because the parameters can be altered to simulate different weather scenarios and facilitate climate change studies [16]. Verdin et al., [15], Furrer and Katz [18], Buishand and Brandsma [33], Seneviratne et al., [34] noted that the non-parametric method has limitations in generating extreme events because values can only be in the range of the observations. Using only the observed sequences ignores climate change's impacts on altering the intensities of the variables and is insufficient in assessing the future response of water resources systems because it leads to single results corresponding only to these observed sequences [22,23,25,34–36].

Most existing SWGs are for single sites and cannot capture the spatial and cross correlations between the variables, which are essential for generating realistic climate change scenarios. Schaake et al., [37] stated that "relationships between physically dependent variables like precipitation and temperature should be respected". Single site SWGs can fail to capture the extreme events of the generated runoff, which are essential to develop realistic adaptation strategies to cope with flood and drought events, especially where a high runoff in one sub-basin can be offset by the low runoff in adjacent sub-basins [26,35,38].

Moreover, the misrepresentation of spatial and cross correlations (e.g., correlations between the precipitation and temperature) leads to biased generated streamflows as this correlation determines the water availability for evapotranspiration and snowmelt [32,39,40]. Therefore, SWGs should capture the characteristics of each site and the spatial dependence among them.

Recently, multi-site and multi-variable SWGs have been developed using different approaches. Steinschneider and Brown, [4] developed a semi-parametric model using a k-nearest-neighbor resampling scheme to simulate multiple spatially distributed variables using wavelet decomposition and autoregressive model to account for low-frequency oscillations. They used a Markov chain of first-order with three states to identify the precipitation states (e.g., dry, wet, and extremely wet). This model had difficulty in preserving the weather statistics besides the cross correlation. Additionally, it is not clear how to diagnose the differences between the precipitation states (e.g., wet and extremely wet).

Srivastav and Simonovic, [32] developed a non-parametric model using the maximum entropy bootstrap technique to capture the time-dependent structure and statistical characteristics. They used an orthogonal transformation to capture the spatial correlations. Even though the model preserves the historical characteristics, Verdin et al., [15] and Chen et al., [40] showed that the maximum entropy bootstrap technique is limited to the historical data range leading to inadequacy in climate change studies. It is difficult to employ this model to create different climate scenarios through variations of parameters.

Li and Babovic, [41] proposed a two-stage parametric model using an empirical copula to generate spatial distribution templates. Then, they developed a rank ordering technique that depended on historic data ranks with an empirical copula technique to preserve the correlations between the variables. The model preserves correlations between the variables and sites but is limited to the historic record length. For example, the model cannot generate more than 30 years of simulation if the historic observations are 30 years. Therefore, the model is not useful in areas with limited data length as an insufficient projection length may lead to wrong conclusions in risk assessment studies [42,43].

Verdin et al., [15] presented a model using a Bayesian hierarchical technique. The precipitation amounts are modeled using gamma distributions and maximum and minimum temperatures are modeled using a normal distribution. The statistical coefficients within them are modeled as spatial Gaussian processes to account for the correlations. Besides the complexity of model structure, the model has difficulty in preserving the statistical properties of the variables (especially the standard deviation of the minimum temperature is extremely underestimated by the model). Additionally, the model underestimates the spatial correlation between the variables. Furthermore, their results do not demonstrate the model's ability to preserve the cross correlation between the variables as well as the temporal correlation.

3. Model Description

The goal here is to develop a parametric regional weather generator (PR-WG) to generate daily stochastic weather variables that preserve their statistical parameters, such as the mean and standard deviations, as well as the spatial, temporal, and cross correlations among them. It should be easy to implement and adapt by altering the statistical parameters to generate synthetic future climate scenarios. The generated scenario must exceed the historic record length and observation range.

The novel contribution is to use a parametric approach to create a flexible model that can adapt to any continuous probability distribution. This will enable the use of the most accurate distribution for each weather variable, and the user can employ other distributions according to the data availability and scope of the study.

3.1. Precipitation States

The first step in developing the PR-WG is to establish the precipitation states. They are defined here as: wet days if the daily amounts equal or exceed 0.1 mm and dry days otherwise. This is similar to the approach by Verdin et al., [15] and Li and V. Babovic [41]. The approach is to use the first-order two-state Markov chain (FTMC), which is the most popular method to produce dry and wet precipitation occurrences. It works well in different climate types and performs as well as higher Markov chain orders [21,22].

Let $S_{(k,t,m)}$ denote the precipitation state ($S = 0$ is a dry day and $S = 1$ is a wet day) at spatial location $k \in \mathbb{N}$, time index $t \in \mathbb{N}$ in days, and month index $m = \{1,2, \dots 12\}$. The dry or wet day occurrence is obtained from the following conditional probabilities:

$$Pr\left(S_{(k,t,m)} = 0 \,|S_{k,t-1,m} = 0\right) = \kappa_0 \,; \, Pr\left(S_{(k,t,m)} = 1 \,|S_{k,t-1,m} = 0\right) = 1 - \kappa_0 \tag{1}$$

$$Pr\left(S_{(k,t,m)} = 1 \,|S_{k,t-1,m} = 1\right) = \kappa_1 \,; \, Pr\left(S_{(k,t,m)} = 0 \,|S_{k,t-1,m} = 1\right) = 1 - \kappa_1 \tag{2}$$

where, κ_0 is the probability of a dry day following a dry day, and κ_1 is the probability of a wet day following a wet day. These probabilities were estimated from the daily historical precipitation observations for each month.

3.2. Precipitation Amount

Precipitation amounts were calculated by using the joint probability distribution between the occurrence and amount. For example, once a wet day is predicted from the FTMC, the precipitation amount is calculated. A skewed normal distribution (SN) was selected because it was recommended by other researchers and estimates the daily precipitation amount better than other distributions such as exponential, gamma, Weibull, mixed-exponential, and generalized Pareto in capturing the mean, standard deviation, and extreme values [20,21,36,44,45].

Let P denote the precipitation amount in mm/day and $\mathbb{1}_\Psi$ denote the indicator of precipitation state condition ψ. P returns to a value obtained implicitly from Equation (4) [46] if the condition ψ holds $(\mathbb{1}_{[S=1]})$ and returns to zero otherwise $(\mathbb{1}_{[S=0]})$, as follows:

$$P_{(k,t,m)} = \begin{cases} SN\left(\mu_P, \sigma_P, \gamma_P\right) & for \quad \mathbb{1}_{[S(k,t,m)=1]} \\ 0 & for \quad \mathbb{1}_{[S(k,t,m)=0]} \end{cases} \tag{3}$$

$$\theta_{(k,t,m)} = \frac{6}{\gamma_{p(k,m)}} \left\{ \left[\frac{\gamma_{p(k,m)}}{2} \left(\frac{P_{(k,t,m)} - \mu_{p(k,m)}}{\sigma_{p(k,m)}} \right) + 1 \right]^{\frac{1}{3}} - 1 \right\} + \frac{\gamma_{p(k,m)}}{6} \tag{4}$$

where θ is the matrix of the standard normal deviates $\theta \sim N(0,1) \in \mathbb{R}$, and μ_p, σ_p, and γ_p, are the mean, standard deviation, and skew coefficient of the precipitation for month m. The values of the parameters μ_p, σ_p, and γ_p were estimated from the daily historical observations using the method of maximum likelihood estimation (MLE).

3.3. Maximum and Minimum Air Temperature

The maximum and minimum daily air temperatures are usually modeled by the normal distribution (N) [47,48]. Let T_X and T_N denote the maximum and minimum daily air temperature in °C, respectively. In which, T_X is (and T_N) is:

$$T_{X(k)} \sim N\left(\mu_{X(k)}, \sigma_{X(k)}\right) \tag{5}$$

where μ_X and σ_X are the mean and standard deviation of T_X, respectively. Solving Equation (5) for each month m according to $\mathbb{1}_\Psi$ (to account for precipitation state effects), T_x and T_N can be computed as:

$$T_{X(k,t,m)} = \mu_{x0(k,m)} + \sigma_{X0(k,m)} \times \mathbb{1}_{(k,t,m)} \; for \; \mathbb{1}_{[S(k,t,m)=0]} \tag{6}$$

$$T_{X(k,t,m)} = \mu_{\mu x1(k,m)} + \sigma_{X1(k,m)} \times \mathbb{1}_{(k,t,m)} \; for \; \mathbb{1}_{[S(k,t,m)=1]} \tag{7}$$

$$T_{N(k,t,m)} = \mu_{\mu N0(k,m)} + \sigma_{N0(k,m)} \times \delta_{(k,t,m)} \; for \; \mathbb{1}_{[S(k,t,m)=0]} \tag{8}$$

$$T_{N(k,t,m)} = \mu_{\mu N1(k,m)} + \sigma_{N1(k,m)} \times \delta_{(k,t,m)} \; for \; \mathbb{1}_{[S(k,t,m)=1]} \tag{9}$$

where, μ_{X0}, μ_{X1}, μ_{N0}, μ_{N1}, σ_{X0}, σ_{X1}, σ_{N0}, and σ_{N1} are the monthly mean and standard deviation for the maximum and minimum air temperature (°C/day) for $S = 0$ and 1, respectively, and v and δ are the matrices of standard normal deviates, such that v and $\delta \sim N(0,1) \in \mathbb{R}$. The parameter values of Equations (6)–(9) were estimated from the historic observations using MLEs.

3.4. Wind Speed Magnitude

Ref. [49] showed that the most accurate function to simulate the daily wind speed magnitude (WS) is Weibull with three and two parameters, respectively, followed by gamma. Given the condition that wind speed is affected by precipitation states and amounts [50], the selected distribution must be decomposed into the same distribution type. As the Weibull distribution cannot be decomposed into two Weibulls (although gamma can be [51]), wind speed magnitude was modeled by the gamma distribution (GM) in this study. Let WS denote the daily wind speed magnitude (m/s) for k locations, as follows:

$$WS_{(k)} \sim GM\left(\alpha_{(k)}, \beta_{(k)}\right) \tag{10}$$

where α and β are the shape and scale parameters, respectively. Similarly for the temperature, the WS for each month m, according to $\mathbb{I}_\Psi$, was estimated implicitly from the following equations:

$$\lambda_{(k,t,m)} = \frac{\beta_{0(k,m)}^{-\alpha_{0(k,m)}}}{\Gamma\left(\alpha_{0(k,m)}\right)} \int_0^{WS_{(k,t,m)}} h^{\alpha_{0(k,m)}-1} \, exp^{-h/\beta_{0(k,m)}} \, dh \; for \; \mathbb{I}_{[S(k,t,m)=0]} \tag{11}$$

$$\lambda_{(k,t,m)} = \frac{\beta_{1(k,m)}^{-\alpha_{0(k,m)}}}{\Gamma\left(\alpha_{1(k,m)}\right)} \int_0^{WS_{(k,t,m)}} h^{\alpha_{1(k,m)}-1} \, exp^{-h/\beta_{1(k,m)}} \, dh \; for \; \mathbb{I}_{[S(k,t,m)=1]} \tag{12}$$

where α_0, α_1, β_0, and β_1 are the shape and scale parameters for $S = 0$ and 1, respectively, for each month m, h is an independent parameter, and λ is the cumulative probability, which is distributed uniformly—$\lambda \sim U[0, 1]$, $\epsilon \, \mathbb{R}$. The shape and scale parameters were estimated from the historic observations using MLEs.

4. Model Implementation

The parametric SWG should conserve the spatial, temporal, and cross correlations of the historic observations of the four weather variables. The concept is to study the behavior of the variates θ, v, δ, and λ, hereafter referred to as anomalies. The correlations between those anomalies should be identified so the generated weather values are statistically similar to the observed values and conserve spatial, temporal, and cross correlations. The implementation of the PR-WG consists of two stages, namely preprocessing and postprocessing, as shown in Figure 1.

4.1. Preprocessing: Parameter Estimation and Matrix Preparation

In order to specify the wet and dry occurrences, a random uniform variate y ~U(0, 1) must be drawn and compared with the transition probabilities obtained from Equations (1) and (2). For multi-site precipitation, the anomalies (referred to as Y ϵ $\mathbb{R}$) that identify the states in k locations must be correlated so that the generated states S are correlated to the historic observations. Wilks' method was selected to generate correlated anomalies Y~ N(0,1) at multiple sites. It is simple and more efficient than hidden Markov and k-nearest neighbor methods [52], accurate in generating the correlations of monthly interstations [53], and the most cited method compared to other approaches [54].

Assume S (1,m) and S (2,m) are the precipitation states on month m at sites $k = 1$ and $k = 2$. To generate realistic sequences of the precipitation states at these two sites, the correlation (ω) between their corresponding anomalies Y, $\omega_{(1,2)} =$ corr $(Y_{(1,m)}, Y_{(2,m)})$ must be computed. The parameter ω was determined by generating different sets of $\acute{Y}$ at the two sites with different arbitrary correlation values $\{\acute{\omega}_1, \acute{\omega}_2, \dots \}$, $\acute{\omega}_1 =$ corr $(\acute{Y}_{(1,m)}, \acute{Y}_{(2,m)})$, identifying the precipitation states at the two locations $\acute{S}_1$ and $\acute{S}_2$, and calculating the corresponding correlation $\{\acute{\varepsilon}_1, \acute{\varepsilon}_2, \dots \}$, $\acute{\varepsilon}_{1(1,2)} =$ corr $(\acute{S}_{(1,m)}, \acute{S}_{(2,m)})$. Then, a regression line between $\acute{\varepsilon}$ and $\acute{\omega}$ sets was fitted to identify the relationship between them. Using this regression equation with the observed precipitation state correlation $\acute{\varepsilon}$, the parameter ω can then be found. A synthetic example is shown in Figure 2a, in which selecting a 0.858 correlation between the pair anomalies (ω) will produce 0.785 correlation between the pair states ($\acute{\varepsilon}$) at the two locations.

The process should be repeated for each station pair and lead to the number of realizations of k (k-1)/2 and be repeated for each month m to create the anomalies matrix $\omega_s \in \mathbb{R}$. The ω_s matrix is then used to develop Y that produces correlated precipitation states in k locations for month m, using the multivariate normal distribution as follows,

$$Y = f\left(\mu_y, \Sigma\right) = \frac{1}{\sqrt{\Sigma \, (2\pi)^d}} exp\left(-\frac{1}{2}\left(y - \mu_y\right)\Sigma^{-1}\left(y - \mu_y\right)\right) \tag{13}$$

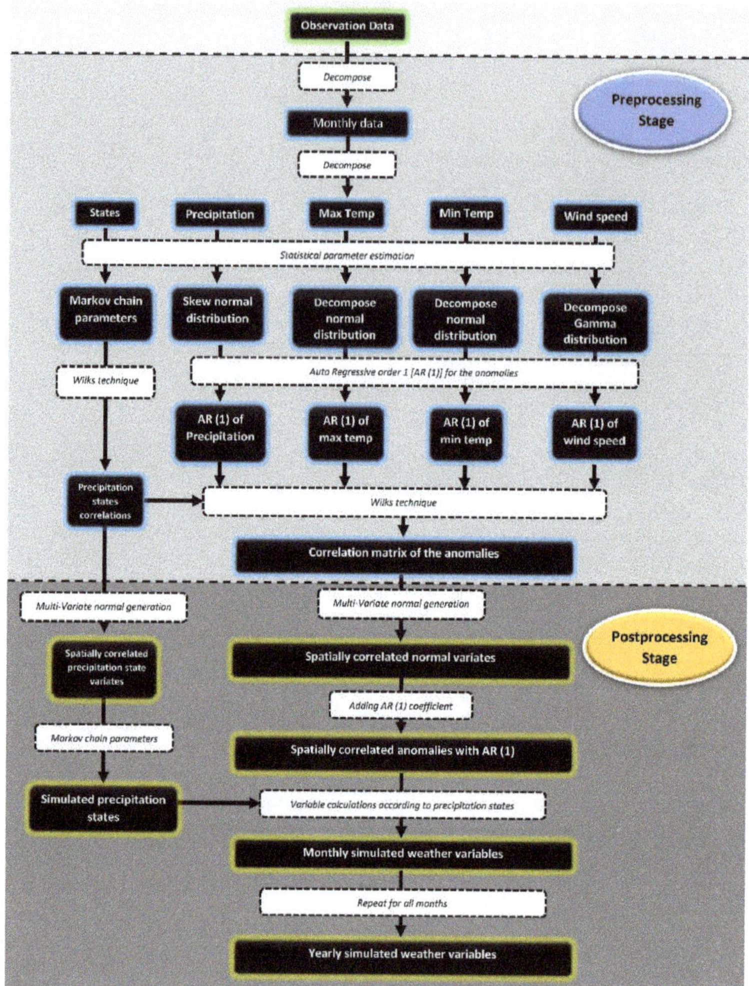

Figure 1. Schematic flowchart of the daily weather generation processes.

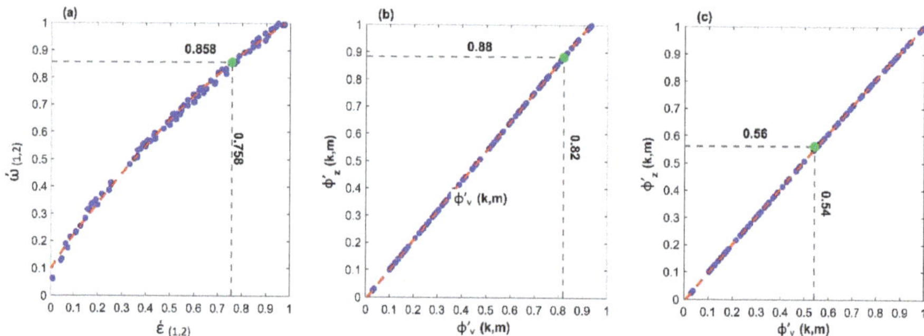

Figure 2. (**a**) An example of Wilks' technique for precipitation states; (**b**) and (**c**) are examples of Wilks' technique to obtain φ_z for T_x and WS, respectively, for station k of month m.

The variable μ_y denotes the 1-D mean vector for the anomalies Y, Σ denotes the covariance matrix, and d is an independent parameter. In this case, $\mu = [0, 0, \ldots, 0]_{k \times 1}$ and the variance is 1, so the covariance matrix Σ_s becomes the correlation matrix ω_s.

The matrix ω_s must be a positive-definite matrix (e.g., the matrix is symmetric and all its eigenvalues are positive) to be implemented in Equation (13). Since the elements of ω_s were calculated empirically, ω_s is usually a non-positive matrix. Comparing to the work of others, the most precise method to obtain a positive-definite matrix is the iterative spectral with Dykstra's correction (ISDC) [55], as follows:

1) Assume $\omega_i = \omega$, $\Delta\Omega_i = 0$, and $i = 1$, in which ω is a non-positive-definite correlation matrix.
2) Let $R_i = \omega_i - \Delta\Omega_i$.
3) Find L_i, and Ω_i, such that $R_i = \Omega_i\, L_i\, \Omega_i^T$.
4) Replace the negative eigenvalues of L_i by a small positive value to construct L_i^+.
5) Set $\omega_{i+1} = \Omega_i\, L_i^+\, \Omega_i^T$ and $\Delta\Omega_{i+1} = \omega_{i+1} - R_i$. Then, replace all ω_{i+1} diagonal elements with 1.
6) Test whether ω_{i+1} is a positive-defined matrix or not. If not, repeat the steps from two to six by making $i = i + 1$ and $\omega_i = \omega_{i-1}$.

After generating the matrix S at k and m, the next step is to simulate the weather variables (e.g., P, T_X, T_N and WS). The idea here is to examine the anomalies of these variables and generate the weather variables with the same observation properties. To account for all the spatial and cross correlation between the variables, their anomalies (θ, υ, δ, and λ) must be correlated. The temporal correlation, identified by the Lag-1 day auto-correlations, for the T_X, T_N and WS must also be considered. Since the precipitation amount is an intermittent variable, the auto-correlation is not considered. The following procedure was suggested to achieve this purpose. First, arrange the weather variable matrix V as follows,

$$
\begin{bmatrix}
V_{1,1}^1 & V_{1,2}^1 & \cdots & V_{1,k}^n \\
V_{2,1}^1 & V_{2,2}^1 & \cdots & V_{2,k}^n \\
\vdots & \vdots & \cdots & \vdots \\
V_{t,1}^1 & V_{t,2}^1 & \cdots & V_{t,k}^n
\end{bmatrix}
\tag{14}
$$

where, V represents the observed weather variable value and n denotes the weather variable rank (P, T_X, T_N and WS), $n = \{1, 2, 3, 4\}$. The total number of the rows will be T = month days × year numbers, the columns will be K × N, and the aisle will be M. This matrix arrangement enables us to consider all the spatial and cross correlations between the weather variables. Next, extract the anomalies matrix Z $\in \mathbb{R}$ from V using Equations (3) and (4) for P; Equations (6)–(9) for T_X and T_N; Equations (11) and (12) for the WS, after estimating their parameters (e.g., μ_p, σ_p, γ_p for P, μ_{X0}, μ_{X1}, σ_{X0}, σ_{X1} for T_X, μ_{N0}, μ_{N1}, σ_{N0}, σ_{N1} for T_N, and α_0, α_1 β_0, β_1 for the WS).

The Z matrix represents the anomalies of the weather variables and their elements have spatial, cross-, and auto-correlation magnitudes. To generate the Z matrix with the same observation properties, these correlations must be preserved. The first step done here was to estimate autoregressive model of order 1, AR(1), coefficients for the anomalies (φ_z) so that the generated variables have the observed AR(1) value (φ_v) applying Wilks' technique. For illustration, synthetically assume that the values of μ_{X0}, μ_{X1}, σ_{X0}, σ_{X1} are 11.72, 9.12, 3.71, 2.21 (C°/day), respectively, and φ_v is 0.82 at station k of month m. The adopted procedure for obtaining the φ_z is as follows:

1) Generate the standard normal random deviate set y; $y \sim N\,(0,1)$.
2) Use y with Equations (1) and (2) to identify the dry and wet days.
3) Generate a standard normal random deviate set x; $x \sim N\,(0,1)$.
4) Apply the AR(1) of arbitrary values between −1 and 1 (e.g., φ'_z).
5) Obtain the anomalies z by standardizing x of Step 4.
6) Apply Equations (6) and (7) to obtain T'_X.

7) Calculate the AR (1) of T_X (e.g., φ'_v) and plot versus the φ'_z, then regress them.
8) Use the regression equation obtained in Step 7 with the observed value φ_v (e.g., 0.82) to determine φ_z. In this case, 0.88 (as shown in Figure 2b).

This procedure must be done for all T_x and T_N of each k and m. For the WS, the procedure is the same except for Step 5, converting x so it is uniformly distributed to get the WS anomalies. For example, let us assume that α_0, α_1, β_0, and β_1 are 4.04, 3.22, 0.62, 0.71, respectively, and the φ_v is 0.54. The corresponding φ_z will be 0.56, as shown in Figure 2c. This procedure allows us to preserve the auto-correlation of T_x, T_N, and the WS.

The final step of the preprocessing stage is to construct the positive-definite correlation matrix of the variable anomalies ω_V, as done for precipitation states using ISDC. Building the ω_V allows us to preserve all the spatial, temporal, and cross correlations between the variables.

4.2. Postprocessing Stage: Variable Generation

After building all matrices and estimating the parameters in the preprocessing stage, the four weather variables can be generated for any time length of interest, as follows:

1) Use Equation (13) with ω_s to generate Y anomalies that denote S. The length of Y denotes the day number of the generated time series. In this case, the user can generate any length (independently of the historic observation length).
2) Use Equations (1) and (2) with the estimated FTMC parameters (κ_0 and κ_1) to identify the dry and wet day occurrences.
3) Apply Equation (13) with ω_v to generate Z anomalies that denote the variable values. Of course, the length of Z must be the same of Y.
4) Obtain P for the wet days using Equations (3) and (4) with the estimated parameters μ_p, σ_p, and ι_p. This will make sure the generated P have similar observed statistics.
5) Apply the AR (1) with coefficients ϕ_z for T_x, T_N and the WS anomalies to consider the auto-correlation magnitude for the variables.
6) Re-standardize the anomalies for T_X and T_N, as follows:

$$Z_{std(k)} = \frac{Z_{(k)} - \mu\left(Z_{(k)}\right)}{\sigma\left(Z_{(k)}\right)} \tag{15}$$

where Z_{std} represents the standardized anomalies Z of Step 5, and $\mu(Z)$ and $\sigma(Z)$ are the mean and standard deviation of Z, respectively.
7) Apply Z_{std} in Equations (6)–(9) with the estimated parameters μ_{X0}, μ_{X1}, μ_{N0}, μ_{N1}, σ_{X0}, σ_{X1}, σ_{N0}, and σ_{N1} to calculate T_x and T_N.
8) Convert the anomalies Z of the WS to be uniformly distributed between 0 and 1 Z_U, as follows:

$$Z_{U(k)} = 0.5 \times erf\left(\frac{Z_{(k)} - \mu\left(Z_{(k)}\right)}{\sqrt{2}\,\sigma\left(Z_{(k)}\right)}\right) + 0.5 \tag{16}$$

9) Apply Z_U in Equations (11) and (12) with the estimated parameters α_0, α_1, β_0, and β_1 to calculate the WS. Steps 3 to 9 enable us to preserve the observation statistics of T_x, T_N and the WS and the spatial, temporal, and cross correlations with consideration of the precipitation states effects through decomposing their distribution functions.
10) Repeat Steps 1 to 9 for all months m.

5. Case Study and Data

The developed PR-WG was tested in the Diyala River basin, which is a transboundary basin between Iran and Iraq with a total stream length of 217 km and basin area of 16,760 km^2 above Derbendikhan Dam, as shown in Figure 3. In previous work, Waheed et al., [5] implemented the daily weather data (e.g., precipitation, maximum and minimum temperature, and wind speed) in this basin at a 0.5° spatial resolution from 1948 to 2006 and explained the implementation procedure. In this follow up study, the historic forcing data were used to validate the proposed PR-WG and test its performance. The reader should refer to the original paper for more details about the data implementation and their validation in the basin.

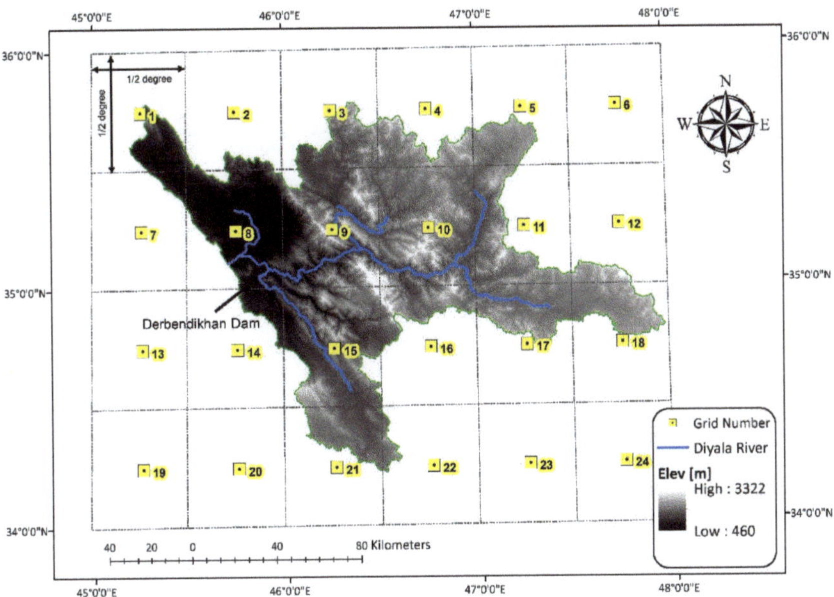

Figure 3. Diyala River basin in Iraq with grid-cell numbers.

6. Results and Discussion

6.1. Model Performance Evaluation

The PR-WG was tested for its daily performance with historic observations for the period between 1948 and 2006, e.g., 58 years, in a grid composed of 24 grid-cells. The Nash–Sutcliffe coefficient efficiency (NSCE; [56]) and the Kling–Gupta efficiency (KGE; [5]) were used to evaluate the PR-WG's ability to produce spatially correlated precipitation states S similar to the observed values, as follows:

$$\text{NSCE} = 1 - \frac{\sum \left(Sim_i - Obs_i \right)^2}{\sum \left(\mu_{sim} - Sim_i \right)^2} \tag{17}$$

$$\text{KGE} = 1 - \sqrt{\left(\frac{\mu_{sim}}{\mu_{obs}} - 1 \right)^2 + \left(\frac{\sigma_{sim}}{\sigma_{obs}} - 1 \right)^2 + (\rho - 1)^2} \tag{18}$$

where *Sim* and *Obs* are the simulations (e.g., the PR-WG outcomes) and the observations of the time index t, respectively; μ_{obs}, σ_{obs}, μ_{sim}, and σ_{sim} are the mean and standard deviation of the observations and simulations (e.g., the PR-WG outcomes), respectively, and ρ is the correlation coefficient between the observations and simulations.

Figure 4 shows the comparison of 10 separate daily simulations each of the same observation length (e.g., 58 years) of PR-WG monthly dry and wet occurrences in gray color dots. The average of these 10 simulations is calculated and plotted in blue dots. The A 1-1 line is also plotted to ease the comparison. It is evident that the model works well to produce the number of dry and wet days, with KGE and NSCE values of 0.97. This result demonstrates the ability of the FTMC to produce the precipitation states well [21,22]. Figure 5 shows a comparison of pairwise correlations of the daily precipitation states calculated for each calendar month. It can be seen that the correlations are captured well by the PR-WG. The overall KGE and NSCE values are 0.98 and 0.99, respectively.

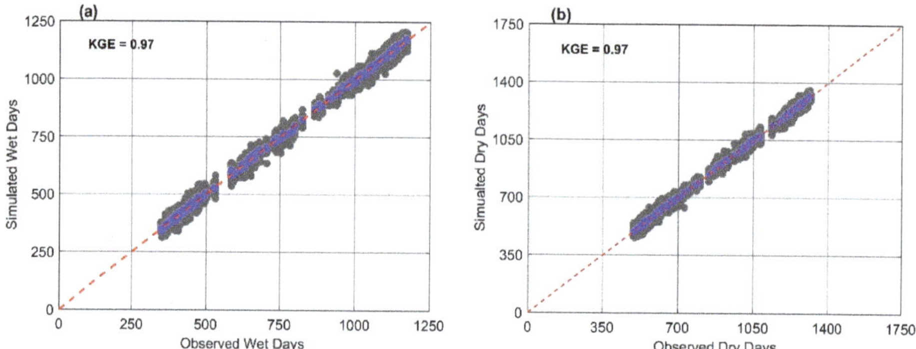

Figure 4. Comparison of the daily precipitation states between the observations and simulations for all months and grid-cells.

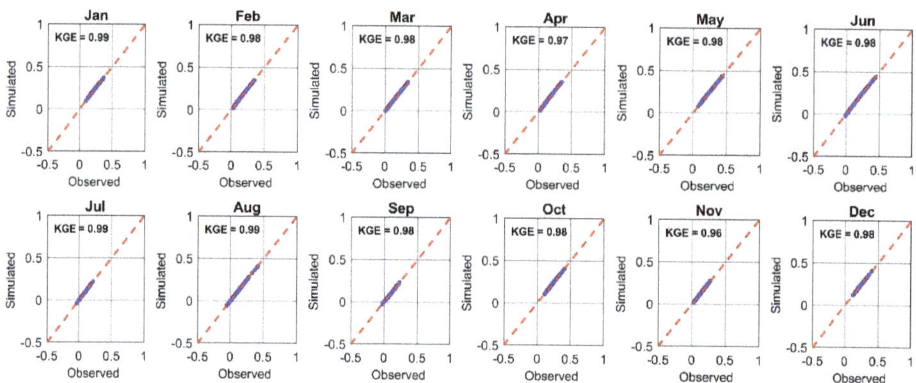

Figure 5. Comparison of the daily precipitation state correlation between the observations and simulations for each month for all grid-cells.

Figure 6 demonstrates the PR-WG performance to produce the statistical parameters (e.g., mean, standard deviation and skewness) of the four weather variables. The comparisons were done on a daily basis at each month for the 24 grid-cells. A daily time step series of 1000 years was generated to reduce the sampling bias and uncertainty in the simulations. However, the daily means of all variables and the standard deviations for T_x, T_N and WS were perfectly produced by the model (KGE $\approx$ 1), while σ_p, and γ_p are reasonably preserved (KGE = 0.96 and 0.86; NSCE = 0.98 and 0.93). The slight discrepancies are due to the stochastic nature of the process [57].

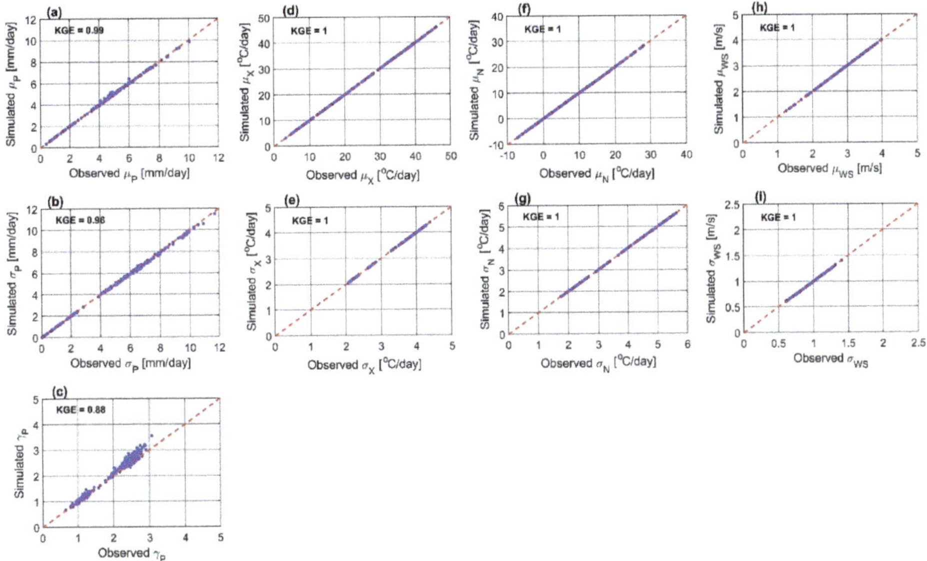

Figure 6. Comparisons of the daily statistic parameters of the observations and simulations. (**a–c**) are the mean, standard deviation, and skewness of P. (**d–g**) are the mean and standard deviation of T_X and T_N. (**h,i**) are the mean and standard deviation of the WS.

Figure 7 shows the daily median values with 0.05 and 0.95 quantiles of the daily values in the bounded areas, and the inverse cumulative distribution function (CDF^{-1}) of the observed and simulated weather variables for grid-cell number 9, which is located in the basin heart (see Figure 3). It is seen that PR-WG well preserves the daily medians for all months. Moreover, the quantile daily estimates show good agreement with the observation quantiles, proving the model's ability to capture the maximum and minimum daily weather values. It is also noticeable from the inverse CDF the observed and simulated weather values are very close, evincing the validity of the selected distribution types. Furthermore, the simulated daily values of quantile 1 exceed the observation values which demonstrates the model's ability to produce values beyond the observation ranges.

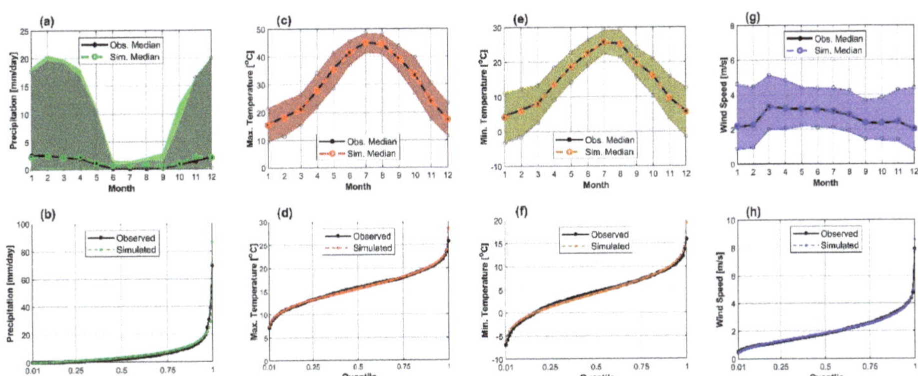

Figure 7. Comparisons of the daily observed and simulated values for the medians with daily 0.05 and 0.95 quantiles in the bounded areas, and the CDF^{-1} for the four weather variables.

Figure 8 shows the spatial and cross correlation coefficient matrices of the observations and simulations for one month (e.g., m=1), while Figure 9 shows the spatial and cross correlation comparison for all variables for each m calculated at daily time steps. The number of columns of the V matrix (see Section 4.1) are $4 \times 24 = 96$. Therefore, the V dimensions are 96×96, in which the values are from 1 to 24 for P, 25 to 48 for T_x, 49 to 72 for T_N, and 73 to 96 for the WS. It can be observed from Figure 8 that the observed correlation among the variables varies greatly across them. P and the WS are slightly less spatially correlated as compared with T_x and T_N. These facts are in line with Srivastav and Simonovic [32] and Verdin et al. [15]. It is also noticeable from Figures 8 and 9 that the model preserves the spatial and cross correlation well among the variables. The overall KGE and NSCE values are 0.96 and 0.97, respectively.

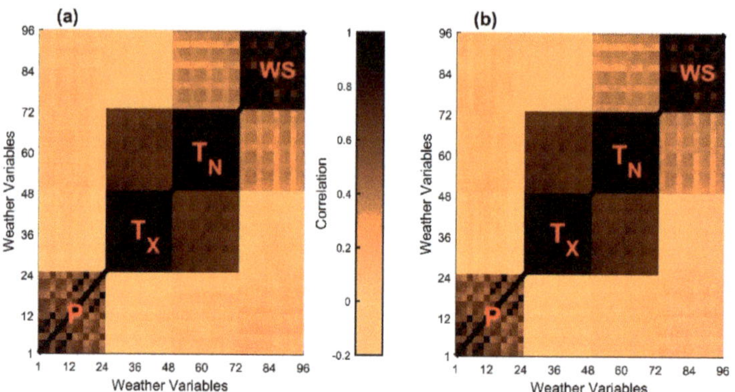

Figure 8. Spatial and cross correlation coefficients of the daily observed (**a**) and simulated variables (**b**).

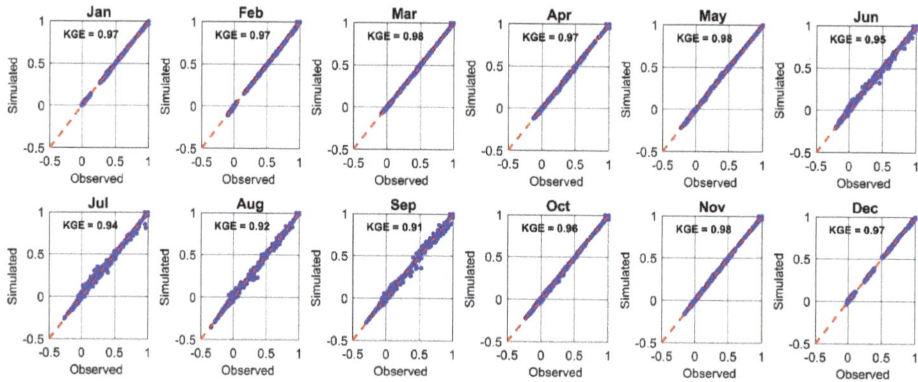

Figure 9. Spatial and cross correlation comparison of the daily weather variables for each month.

Figure 10 demonstrates the PR-WG capability to preserve the Lag-1 day auto-correlations of T_x, T_N and the WS. It is noticeable that the values differ from month to month, they are less for the WS comparing to T_x and T_N. However, the PR-WG captures these monthly variations very well regardless of their magnitudes with the overall KGE and NSCE values of 0.97 and 0.98, respectively.

The results presented here glimpse the model capability to preserve the statistical properties of the observations to synthesize the future scenarios. The proposed model demonstrated the Wilks technique ability to generate anomalies similarly to the observations. It is also seen that the hybrid structure of the AR and Wilks technique leads to generate data that preserve the temporal correlation beside the spatial and cross correlations.

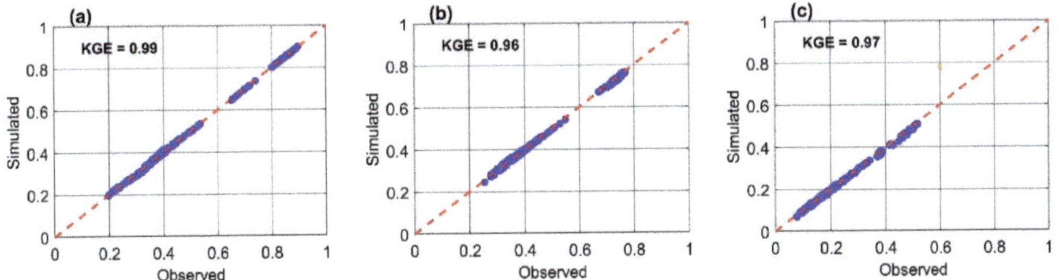

Figure 10. Lag-1 day auto-correlations of the weather variables T_x, T_N, and WS, respectively, for all months.

The key advantage of PR-WG is that it is built to be a general model through studying the observation anomalies and mimicking them. Therefore, the model is anticipated to work well in different climate zones and topographies regardless of the data spatial and temporal scale. The model framework is flexible enough for locations observe short-term and long-term variations. Moreover, the user can reduce the cycle data length to meet their scope. e.g., they can use a data window of two weeks (or a week) instead of the monthly window that was used in this study. The computational expensive of implemented the pre-processing stage has to carefully examined.

6.2. Model Validation

In some cases, the proposed SWG produces negative daily values for precipitation. [58] indicated that the SN is not suitable when the skewness is greater than 4.5. However, in the study area, values of the skewness have not exceeded 4.5 (see Figure 6c), therefore the SN is applicable. The negatives of the daily values were checked and found to be less than 3% of the whole 1000-year time series in the 24 grid-cells. The suggestion of [32] to round the negative values to zero was considered, but it would affect the number of wet and dry calculations and the statistical parameters of precipitation. Instead, the negatives were rounded to 0.1 mm/day, which is assumed to be the minimum precipitation amount (see Section 3.1). This correction approach for negative values illustrates the slight differences in the simulated σ_p and γ_p (see Figure 5b,c). The user could apply another distribution function in cases where the SN is not applicable such as mixed-exponential [59,60], log-normal, gamma, etc. The key advantage of PR-WG is its flexibility in adopting any distribution of interest, such as these.

The second validation was done by checking if T_N is greater than T_x and was found to be less than 1% of the whole 1000-year time series in the 24 grid-cells. [41] suggested to force T_x to be greater than T_N through setting T_N equal to T_x minus 1. This procedure will affect the auto-correlation of the T_N. Instead, the Chen et al., [57] approach was applied as follows, if $T_x < T_N$,

$$T_{X(k,t,m)} = T_{N(k,t,m)} + \left(\mu_{\mu x(k,m)} - \mu_{\mu N(k,m)}\right) + \sqrt{\sigma^2_{X(k,m)} - \sigma^2_{N(k,m)}} \times z_{std(k,t,m)} \quad for \ \mathbb{1}_{[\sigma X_{(k,t,m)} \geq \sigma N_{(k,t,m)}]} \quad (19)$$

$$T_{X(k,t,m)} = T_{N(k,t,m)} + \left(\mu_{\mu x(k,m)} - \mu_{\mu N(k,m)}\right) + \sqrt{\sigma^2_{X(k,m)} - \sigma^2_{N(k,m)}} \times z_{std(k,t,m)} \ for \ \mathbb{1}_{[\sigma X_{(k,t,m)} < \sigma N_{(k,t,m)}]} \quad (20)$$

Equations (19) and (20) are conditioned on the precipitation states. For example, the σ and μ will turn to condition 0 if $S = 0$. In this case, the T_X is guaranteed to be greater than T_N and the auto, spatial, and cross correlations are preserved since they are multiplied by the anomalies z_{std}.

6.3. Model Comparison

For comparison purposes, two SWG models were selected to compare their performances with the PR-WG to further demonstrate the model applicability. The first model is the single site weather generator (WG) developed by Chen et al., [57] and Chen et al., [61]. The second is the two stages weather generator using an empirical copula (EC) approach developed by Li and Babovic [41]. To highlight the unique contribution of the PR-WG model, we focused on the model performances to maintain the spatial, cross, and temporal correlations. Figure 11 shows the daily performances of the WG and EC to for the spatial and cross correlations, while Figure 12 shows the temporal correlations for the sites and month. It is seen that the EC model works well in preserving the spatial, cross, and temporal correlation; the PR-WG is slightly superior to it. However, the KGE and NSCE for the spatial and cross correlations are 0.92 and 0.93, and for the temporal 0.95, and 0.96. It also notable that the WG model poorly preserves the spatial and cross correlations but has good ability to preserve the temporal correlation. This is because the model accounts for the temporal correlation only, where the simulated data were generated independently for all variables and sites which leads to poor spatial and cross correlation accuracy. The KGE and NSCE for the spatial and cross correlations are −0.29 and −8.3, and for the temporal 0.88, and 0.89. Although the EC approach works well in general, the only drawback is that its simulation time period must be identical to the historic observation, which prevents its usage in areas with limited data availability. This is because the post processing stage of the EC model employs a re-ranking technique that extracts the ranked variables directly from the historic observations. Therefore, the model length can only be the same as the historic observations, leading to less flexibility for future scenarios, especially in data scarce regions. The PR-WG has the advantage of producing the simulation length of interest, making it useful in areas with limited data availability besides maintaining the statistical characteristics.

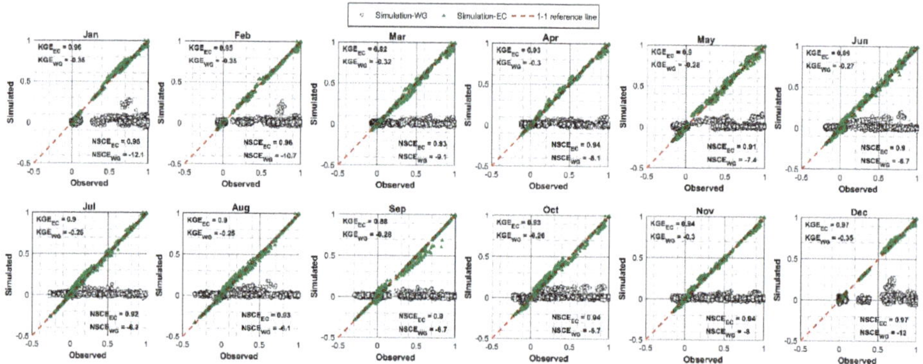

Figure 11. Performance evaluation for empirical copula (EC) and weather generator (WG) models for preserving the spatial and cross correlations of the weather variables for each month.

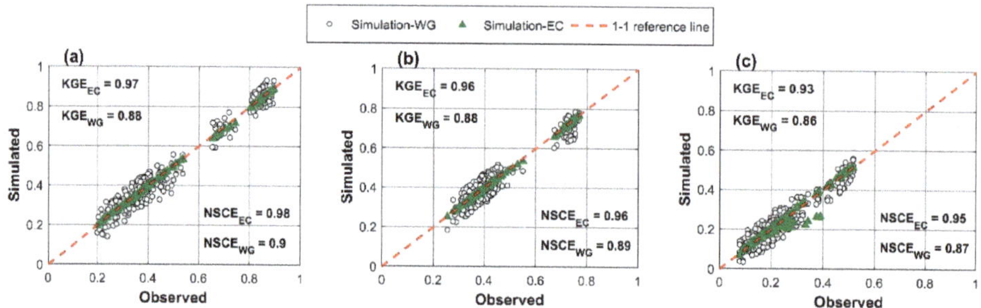

Figure 12. Monthly performance evaluation for EC and WG models for preserving the temporal correlation (Lag-1 day) of the weather variables.

6.4. Simulation of the Future Forecasting Scenarios

The goal of the PR-WG is to be used later for climate variation assessments. The advantage of the model, besides the ability to preserve the statistical characteristics, is its flexibility to alter them to produce a wide range of different scenarios. However, defining the future scenario ranges to test a water resources system's performance in terms of the climate stress is a difficult task and dependent on many factors, including expert opinions [4].

These future scenarios will be applied in the Diyala River basin to discover the vulnerability of the Derbendikhan Dam and its reservoir. Moreover, different adaptation strategies will be suggested in order to test their capabilities to improve the system performance. Since the model is implemented on a stochastic basis, the future trend insights will be obtained from analyzing the GCM models. Then, this can be fed into the PR-WR to mimic the future trend as well as the statistical properties. For instance, multiplicative factors for the precipitation mean will be applied starting from a 0% change in the historical precipitation and annual linearly increasing (or decreasing) up to the specified value in the final period (e.g., +30% of the historical value). Forms other than the linear change can also be applied to synthesize the future forecast data.

7. Conclusions

It was shown that a PR-WG accurately preserves the statistical properties (mean, standard deviation, and skewness coefficient) of the weather variables (overall KGE and NSCE test values were 0.98). The PR-WG also preserves the spatial, temporal, and cross correlations among the weather variables. While other SWGs may have more features, the one developed in this study enables a bottom-up vulnerability assessment study to be implemented in areas with limited data availability.

The PR-WG effectively estimates the dry and wet day occurrences using a FTMC with overall KGE and NSCE values of 0.97, a result that is in line with those in [21,22]. The results also demonstrate the effectiveness of Wilks' technique to produce spatially correlated precipitation states (KGE of 0.98; NSCE of 0.99) and spatially and cross correlated weather variables (KGE of 0.96; NSCE of 0.97), as well as temporally correlated variables (KGE of 0.97; NSCE of 0.98). The model is also capable of preserving the maximum and minimum daily weather values as well as producing values beyond the observed ranges. Furthermore, the PR-WG outperforms the EC and WG models in preserving the spatial, cross, and temporal correlations in the meteorological stations.

While the PR-WG was validated in the Diyala River basin, it should be effective and applicable in other places and with other weather variables, such as solar radiation. The advantages of PR-WG are its flexibility to select any distribution function for each weather variable, ability to simulate any number of years within or beyond the historic observation length, capability to generate values outside the observation range, and its ability to produce synthetic scenarios through the alteration of the weather variable parameters for the study of climate change's impacts. The PR-WG is easy to construct and understand with little computational intensity to build the spatial and cross correlation matrices of the anomalies. Increasing computational power will facilitate the work.

Author Contributions: Conceptualization, S.Q.W., J.A.R., and N.S.G.; methodology, S.Q.W.; software, S.Q.W.; validation, S.Q.W.; formal analysis, S.Q.W., N.S.G., and J.A.R.; investigation, S.Q.W.; resources, S.Q.W.; data curation, S.Q.W.; writing—original draft preparation, S.Q.W.; writing—review and editing, S.Q.W., N.S.G., and J.A.R.; visualization, S.Q.W.; supervision, J.A.R. and N.S.G. All authors have read and agreed to the published version of the manuscript.

Funding: This research was funded by the Iraq Higher Committee for Education Development (HCED), grant number D1201077.

Acknowledgments: The authors are grateful to the Iraqi Ministry of Water Resources for assistance. The authors are also thankful to Xin Li and Vladan Babovic for their cooperation in implementing the EC model. The authors are also grateful to Colorado State University for providing its laboratories and supercomputer to run the model(s) and perform the analyses.

Conflicts of Interest: The authors declare no conflict of interest.

References

1. Hallegatte, S.; Shah, A.; Lempert, R.; Brown, C.; Gill, S. *Investment Decision Making under Deep Uncertainty-Application to Climate Change*; The World Bank: Washington, DC, USA, 2012.

2. Brown, C.; Wilby, R.L. An alternate approach to assessing climate risks. *Eos, Trans. Am. Geophys. Union* **2012**, *93*, 401–402. [CrossRef]

3. Stephenson, D.; Collins, M.; Rougier, J.C.; Chandler, R.E. Statistical problems in the probabilistic prediction of climate change. *Environmetrics* **2012**, *23*, 364–372. [CrossRef]

4. Steinschneider, S.; Brown, C. A semiparametric multivariate, multisite weather generator with low-frequency variability for use in climate risk assessments. *Water Resour. Res.* **2013**, *49*, 7205–7220. [CrossRef]

5. Waheed, S.Q.; Grigg, N.S.; Ramirez, J.A. Variable Infiltration-Capacity Model Sensitivity, Parameter Uncertainty, and Data Augmentation for the Diyala River Basin in Iraq. *J. Hydrol. Eng.* **2020**, *25*, 04020040. [CrossRef]

6. Culley, S.; Noble, S.; Yates, A.; Timbs, M.; Westra, S.; Maier, H.R.; Giuliani, M.; Castelletti, A. A bottom-up approach to identifying the maximum operational adaptive capacity of water resource systems to a changing climate. *Water Resour. Res.* **2016**, *52*, 6751–6768. [CrossRef]

7. Weaver, C.P.; Lempert, R.J.; Brown, C.; Hall, J.A.; Revell, D.; Sarewitz, D. Improving the contribution of climate model information to decision making: The value and demands of robust decision frameworks. *Wiley Interdiscip. Rev. Clim. Chang.* **2012**, *4*, 39–60. [CrossRef]

8. Turner, S.W.; Marlow, D.; Ekström, M.; Rhodes, B.G.; Kularathna, U.; Jeffrey, P. Linking climate projections to performance: A yield-based decision scaling assessment of a large urban water resources system. *Water Resour. Res.* **2014**, *50*, 3553–3567. [CrossRef]

9. Steinschneider, S.; Wi, S.; Brown, C. The integrated effects of climate and hydrologic uncertainty on future flood risk assessments. *Hydrol. Process.* **2014**, *29*, 2823–2839. [CrossRef]

10. Zhang, E.; Yin, X.; Xu, Z.; Yang, Z. Bottom-up quantification of inter-basin water transfer vulnerability to climate change. *Ecol. Indic.* **2018**, *92*, 195–206. [CrossRef]

11. Whateley, S.; Steinschneider, S.; Brown, C. A climate change range-based method for estimating robustness for water resources supply. *Water Resour. Res.* **2014**, *50*, 8944–8961. [CrossRef]

12. Moody, P.; Brown, C. Robustness indicators for evaluation under climate change: Application to the upper Great Lakes. *Water Resour. Res.* **2013**, *49*, 3576–3588. [CrossRef]

13. Steinschneider, S.; McCrary, R.; Wi, S.; Mulligan, K.B.; Mearns, L.O.; Brown, C. Expanded Decision-Scaling Framework to Select Robust Long-Term Water-System Plans under Hydroclimatic Uncertainties. *J. Water Resour. Plan. Manag.* **2015**, *141*, 04015023. [CrossRef]

14. Wilks, D. Multisite generalization of a daily stochastic precipitation generation model. *J. Hydrol.* **1998**, *210*, 178–191. [CrossRef]

15. Verdin, A.P.; Rajagopalan, B.; Kleiber, W.; Podestá, G.; Bert, F. BayGEN: A Bayesian Space-Time Stochastic Weather Generator. *Water Resour. Res.* **2019**, *55*, 2900–2915. [CrossRef]

16. Wilks, D.S. A gridded multisite weather generator and synchronization to observed weather data. *Water Resour. Res.* **2009**, *45*. [CrossRef]

17. Wilks, D.S. *Statistical Methods in the Atmospheric Sciences*; Academic Press: Cambridge, MA, USA, 2011.

18. Furrer, E.; Katz, R.W. Improving the simulation of extreme precipitation events by stochastic weather generators. *Water Resour. Res.* **2008**, *44*. [CrossRef]

19. Jie, C.; Brissette, F.P.; Zhang, X.J. A multi-site stochastic weather generator for daily precipitation and temperature. *Trans. ASABE* **2014**, *57*, 1375–1391.

20. Chen, J.; Brissette, F.P. Stochastic generation of daily precipitation amounts: Review and evaluation of different models. *Clim. Res.* **2014**, *59*, 189–206. [CrossRef]

21. Chen, J.; Brissette, F.P. Comparison of five stochastic weather generators in simulating daily precipitation and temperature for the Loess Plateau of China. *Int. J. Climatol.* **2014**, *34*, 3089–3105. [CrossRef]

22. Acharya, N.; Frei, A.; Chen, J.; DeCristofaro, L.; Owens, E.M. Evaluating Stochastic Precipitation Generators for Climate Change Impact Studies of New York City's Primary Water Supply. *J. Hydrometeorol.* **2017**, *18*, 879–896. [CrossRef]

23. Mukundan, R.; Acharya, N.; Gelda, R.K.; Frei, A.; Owens, E.M. Modeling streamflow sensitivity to climate change in New York City water supply streams using a stochastic weather generator. *J. Hydrol. Reg. Stud.* **2019**, *21*, 147–158. [CrossRef]

24. Mehrotra, R.; Westra, S.; Sharma, A.; Srikanthan, R. Continuous rainfall simulation: 2. A regionalized daily rainfall generation approach. *Water Resour. Res.* **2012**, *48*, 48. [CrossRef]

25. Richardson, C.W. Stochastic simulation of daily precipitation, temperature, and solar radiation. *Water Resour. Res.* **1981**, *17*, 182–190. [CrossRef]

26. Qian, B.; Xu, H. Multisite stochastic weather models for impact studies. *Int. J. Clim.* **2002**, *22*, 1377–1397. [CrossRef]

27. Brissette, F.; Khalili, M.; Leconte, R. Efficient stochastic generation of multi-site synthetic precipitation data. *J. Hydrol.* **2007**, *345*, 121–133. [CrossRef]

28. Srikanthan, R.; Pegram, G. A nested multisite daily rainfall stochastic generation model. *J. Hydrol.* **2009**, *371*, 142–153. [CrossRef]

29. Baigorria, G.A.; Jones, J.W. GiST: A Stochastic Model for Generating Spatially and Temporally Correlated Daily Rainfall Data. *J. Clim.* **2010**, *23*, 5990–6008. [CrossRef]

30. Leander, R.; Buishand, T.A. A daily weather generator based on a two-stage resampling algorithm. *J. Hydrol.* **2009**, *374*, 185–195. [CrossRef]

31. King, L.M.; McLeod, A.I.; Simonovic, S.P. Improved Weather Generator Algorithm for Multisite Simulation of Precipitation and Temperature. *JAWRA J. Am. Water Resour. Assoc.* **2015**, *51*, 1305–1320. [CrossRef]

32. Srivastav, R.K.; Simonovic, S.P. Multi-site, multivariate weather generator using maximum entropy bootstrap. *Clim. Dyn.* **2014**, *44*, 3431–3448. [CrossRef]

33. Khalili, M.; Brissette, F.; Leconte, R. Effectiveness of Multi-Site Weather Generator for Hydrological Modeling1. *JAWRA J. Am. Water Resour. Assoc.* **2011**, *47*, 303–314. [CrossRef]

34. Murray, V.; Ebi, K.L. IPCC Special Report on Managing the Risks of Extreme Events and Disasters to Advance Climate Change Adaptation (SREX). *J. Epidemiol. Community Heal.* **2012**, *66*, 759–760. [CrossRef] [PubMed]

35. Mehrotra, R.; Li, J.; Westra, S.; Sharma, A. A programming tool to generate multi-site daily rainfall using a two-stage semi parametric model. *Environ. Model. Softw.* **2015**, *63*, 230–239. [CrossRef]

36. Wang, W.; Flanagan, D.C.; Yin, S.; Yu, B. Assessment of CLIGEN precipitation and storm pattern generation in China. *Catena* **2018**, *169*, 96–106. [CrossRef]

37. John, S.; Pailleux, J.; Thielen, J.; Arritt, R.; Hamill, T.; Luo, L.; Martin, E.; McCollor, D.; Pappenberger, F. Summary of recommendations of the first workshop on Postprocessing and Downscaling Atmospheric Forecasts for Hydrologic Applications held at Météo-France, Toulouse, France, 15–18 June 2009. *Atmos. Sci. Lett.* **2010**, *11*, 59–63.

38. Li, Z. A new framework for multi-site weather generator: A two-stage model combining a parametric method with a distribution-free shuffle procedure. *Clim. Dyn.* **2013**, *43*, 657–669. [CrossRef]

39. Li, C.; Sinha, E.; Horton, D.E.; Diffenbaugh, N.S.; Michalak, A.M. Joint bias correction of temperature and precipitation in climate model simulations. *J. Geophys. Res. Atmos.* **2014**, *119*, 13–153. [CrossRef]

40. Chen, J.; Li, C.; Brissette, F.P.; Chen, H.; Wang, M.; Essou, G.R. Impacts of correcting the inter-variable correlation of climate model outputs on hydrological modeling. *J. Hydrol.* **2018**, *560*, 326–341. [CrossRef]

41. Li, X.; Babovic, V. A new scheme for multivariate, multisite weather generator with inter-variable, inter-site dependence and inter-annual variability based on empirical copula approach. *Clim. Dyn.* **2018**, *52*, 2247–2267. [CrossRef]

42. Guillermo, A.B.; Jones, J.W. GiST: A stochastic model for generating spatially and temporally correlated daily Investment Decision Making under Deep Uncertainty—Application to Climate Change rainfall data. What kind of data is needed to identify climate impacts? How can data be managed and organized through data catalogues? *J. Clim.* **2010**, *23*, 5990–6008.

43. Haugen, A.; Bertolin, C.; Leijonhufvud, G.; Olstad, T.; Broström, T. A Methodology for Long-Term Monitoring of Climate Change Impacts on Historic Buildings. *Geosciences* **2018**, *8*, 370. [CrossRef]

44. Li, Z.; Brissette, F.; Chen, J. Assessing the applicability of six precipitation probability distribution models on the Loess Plateau of China. *Int. J. Clim.* **2013**, *34*, 462–471. [CrossRef]

45. Mehan, S.; Guo, T.; Gitau, M.W.; Flanagan, D.C. Comparative Study of Different Stochastic Weather Generators for Long-Term Climate Data Simulation. *Climate* **2017**, *5*, 26. [CrossRef]

46. Nicks, A.D.; Gander, G.A. CLIGEN: A Weather Generator for Climate Inputs to Water Resource and Other Models. In Proceedings of the Fifth International Conference on Computers in Agriculture, Orlando, FL, USA, 6–9 February 1994. Available online: https://www.worldcat.org/title/cligen-a-weather-generator-for-climate-inputs-to-water-resource-and-other-models/oclc/693437629 (accessed on 7 August 2020).

47. Harmel, R.D.; Richardson, C.W.; Hanson, C.L.; Johnson, G.L. Evaluating the Adequacy of Simulating Maximum and Minimum Daily Air Temperature with the Normal Distribution. *J. Appl. Meteorol.* **2002**, *41*, 744–753. [CrossRef]

48. Harmel, R.D.; Richardson, C.W.; Hanson, C.L.; Johnson, G.L. Simulating maximum and minimum daily temperature with the normal distribution. In Proceedings of the 2001 ASAE Annual Meeting. American Society of Agricultural and Biological Engineers, Sacramento, CA, USA, 29 July–1 August 2001.

49. Pobočíková, I.; Sedliačková, Z.; Michalková, M. Application of Four Probability Distributions for Wind Speed Modeling. *Procedia Eng.* **2017**, *192*, 713–718. [CrossRef]

50. Back, L.E.; Bretherton, C.S. The Relationship between Wind Speed and Precipitation in the Pacific ITCZ. *J. Clim.* **2005**, *18*, 4317–4328. [CrossRef]

51. Saralees, N. A review of results on sums of random variables. *Acta Appl. Math.* **2008**, *103*, 131–140.

52. Mehrotra, R.; Srikanthan, R.; Sharma, A. A comparison of three stochastic multi-site precipitation occurrence generators. *J. Hydrol.* **2006**, *331*, 280–292. [CrossRef]

53. Khalili, M.; Leconte, R.; Brissette, F. Stochastic Multisite Generation of Daily Precipitation Data Using Spatial Autocorrelation. *J. Hydrometeorol.* **2007**, *8*, 396–412. [CrossRef]

54. Chen, J.; Brissette, F.P.; Zhang, X. Hydrological Modeling Using a Multisite Stochastic Weather Generator. *J. Hydrol. Eng.* **2016**, *21*, 04015060. [CrossRef]

55. Maree, S.C. Correcting Non Positive Definite Correlation Matrices. Bachelor's Thesis, Department of Applied Mathematics, Delft University of Technology, Delft, Australia, 2012.

56. Eamonn, N.J.; Sutcliffe, J.V. River flow forecasting through conceptual models part I—A discussion of principles. *J. Hydrol.* **1970**, *10*, 282–290.

57. Chen, J.; Brissette, F.P.; Leconte, R.; Caron, A. A Versatile Weather Generator for Daily Precipitation and Temperature. *Trans. ASABE* **2012**, *55*, 895–906. [CrossRef]

58. Meyer, C. *General Description of the CLIGEN Model and Its History*; USDA-ARS National Soil Erosion Laboratory: West Lafayette, IN, USA, 2011.

59. Roldań, J.; Woolhiser, D.A.; Roldán, J. Stochastic daily precipitation models: 1. A comparison of occurrence processes. *Water Resour. Res.* **1982**, *18*, 1451–1459. [CrossRef]

60. Wilks, D.S. Simultaneous stochastic simulation of daily precipitation, temperature and solar radiation at multiple sites in complex terrain. *Agric. For. Meteorol.* **1999**, *96*, 85–101. [CrossRef]

61. Chen, J.; Brissette, F.; Leconte, R. A daily stochastic weather generator for preserving low-frequency of climate variability. *J. Hydrol.* **2010**, *388*, 480–490. [CrossRef]

MDPI

St. Alban-Anlage 66

4052 Basel

Switzerland

Tel. +41 61 683 77 34

Fax +41 61 302 89 18

www.mdpi.com

Climate Editorial Office

E-mail: climate@mdpi.com

www.mdpi.com/journal/climate